SpringerBriefs in Food, Health, and Nutrition

T0039135

Editor-in-Chief
Richard W. Hartel
University of Wisconsin – Madison, USA

Associate Editors
J. Peter Clark, *Consultant to the Process Industries, USA*
John W. Finley, *Louisiana State University, USA*
David Rodriguez-Lazaro, *ITACyL, Spain*
Yrjo Roos, *University College Cork, Ireland*
David Topping, *CSIRO, Australia*

Springer Briefs in Food, Health, and Nutrition present concise summaries of cutting edge research and practical applications across a wide range of topics related to the field of food science, including its impact and relationship to health and nutrition. Subjects include:

- Food Chemistry, including analytical methods; ingredient functionality; physic-chemical aspects; thermodynamics
- Food Microbiology, including food safety; fermentation; foodborne pathogens; detection methods
- Food Process Engineering, including unit operations; mass transfer; heating, chilling and freezing; thermal and non-thermal processing, new technologies
- Food Physics, including material science; rheology, chewing/mastication
- Food Policy
- And applications to:
  - Sensory Science
  - Packaging
  - Food Quality
  - Product Development

We are especially interested in how these areas impact or are related to health and nutrition.

Featuring compact volumes of 50 to 125 pages, the series covers a range of content from professional to academic. Typical topics might include:

- A timely report of state-of-the art analytical techniques
- A bridge between new research results, as published in journal articles, and a contextual literature review
- A snapshot of a hot or emerging topic
- An in-depth case study
- A presentation of core concepts that students must understand in order to make independent contributions

For further volumes:
http://www.springer.com/series/10203

Q. Tuan Pham

Food Freezing and Thawing Calculations

Springer

Q. Tuan Pham
School of Chemical Engineering
UNSW Australia
Sydney
New South Wales
Australia

ISSN 2197-571X   ISSN 2197-5728 (electronic)
ISBN 978-1-4939-0556-0   ISBN 978-1-4939-0557-7 (eBook)
DOI 10.1007/978-1-4939-0557-7
Springer New York Heidelberg Dordrecht London

Library of Congress Control Number: 2014933269

Springer is part of Springer Science+Business Media (www.springer.com)

Preface

This work is intended mainly as an introduction to the field for specialists and researchers in food freezing. It may also contain some new or recent materials that would be of interest to more experienced researchers, while industry practitioners may gain from it important insights that would help them solve practical problems or optimise their processes. The first five chapters, which cover property estimation and heat transfer calculations, are sufficiently detailed to enable the reader to carry out these calculations with confidence and get a full understanding of the assumptions involved. Chapter 6, which reviews the modelling of heat transfer coupled with other processes, is necessarily less detailed due to the wide scope of coverage. In this chapter, rather than providing a comprehensive but superficial review, the main lines of approach for each topic are described, together with illustrative examples, so that the reader can get a feel of the methodologies.

Most of the content in this work is from previous literature, but where it is felt that improvements could be made or the development could be more rigorous, new material has been introduced. This is the case for dynamic heat load calculations, heat and mass transfer in porous materials and calorimetric properties and freezing calculations at high pressure.

Freezing occupies a special place in food preservation technology as it combines the benefits of long shelf life—months or years—with excellent retention of nutrients and sensory qualities and complete absence of microbial growth. With the advent of industrial food freezing, which began when frozen meat started being shipped to Europe from southern hemisphere countries (Argentina in 1877, Australia in 1879, New Zealand in 1882), there came a demand for scientifically based design and calculation techniques that would maximize the economic benefits. The accurate prediction of food freezing time is essential for the design of efficient freezing equipment and processes, since over-design is expensive and under-design, which may cause health risks and product losses, is even more so.

For decades, practically the only generic method available for calculating freezing time was Plank's equation, presented in 1913, which gives significant under-prediction. Over the years, many product-specific empirical equations were developed, until Cleland and Earle presented their generic empirical equation, still inspired by Plank's theoretical one, in 1977. In subsequent years, a large number of freezing time formulas was proposed. Simultaneously, computers became widely available and numerical methods such as finite difference and finite element were widely applied. The rigorousness and flexibility of the numerical methods allow foods with complex structures, properties and geometries to be modelled to any required precision. Their predictions were at first regarded with suspicion by industry practitioners and even some researchers, who were more used to empiricism than computer modelling and were often distrusting of "theory". Over the years, however, with rapidly evolving computer power and modelling techniques, many physical phenomena could be accurately modelled and numerical predictions are now generally regarded as reliable, at least when the underlying physical phenomena are well understood and their effect rigorously modelled (one case where these conditions are still not fulfilled is computational fluid dynamics, where turbulent flow must still be modelled with the aid of highly empirical relationships).

In 1990, Andrew Cleland wrote his book *Food Refrigeration Processes Analysis, Design and Simulation*, which summarised progress in the development of analytical, approximate, empirical and numerical methods for food refrigeration calculations to that date. Since then, there have been many other developments in the field. While the basic methodology of numerical modelling remain the same, more accurate data and models of thermophysical properties of food have become available, improving the accuracy of the numerical predictions. The increase in computer power, improvements in modelling techniques and wide availability of commercial numerical packages have allowed phenomena such as mass transfer, ice nucleation, crystal growth and mechanical effects to be modelled alongside heat transfer. In the future, more attention may be paid to multi-scale modelling including the modelling of crystallisation and mass transfer on the microscopic or cellular level, which have important effects on food quality.

It must be said, however, that one can still easily put too much trust in the computer, as no amount of computer power or numerical technique can compensate for a faulty understanding of physical phenomena or incorrect input data. On the contrary, computers can amplify the effects of conceptual shortcomings. In many cases, simple analytical, approximate or empirical methods still have their use in modelling and prediction. They can also quickly provide insights into the physical phenomena or guidance for potential process improvements. Whatever the method used, the best results can only be obtained if its underlying theory, assumptions and limitations are thoroughly understood. This work has been written with this point in mind: the emphasis is not so much on comprehensiveness, although an attempt has been made to provide balanced coverage, than on highlighting the physical principles and reasoning behind the methods considered, so that the reader can apply them judiciously and get a reasonable idea of their reliability.

The author is grateful to Professor Donald Cleland for suggesting this book and to Don and Dr. Simon Lovatt for their helpful comments and suggestions. He would also like to acknowledge many colleagues, particularly Professor Andrew Cleland, for the help, friendship and intellectual stimulation they have given over the years.

Contents

Nomenclature

Acronyms

b.c.	Boundary condition
CFD	Computational fluid dynamics
EHTD	Equivalent heat transfer dimensionality (shape factor)
FDM	Finite difference method
FEM	Finite element method
FVM	Finite volume method
MCP	Mean conducting path
ODE	Ordinary differential equation
PDE	Partial differential equation

Symbols

A	Surface area, m^2
a_w	Water activity
b	Bound water to solids mass ratio, X_b/X_{ds}
Bi	Biot number, hR/k
c	Specific heat (may or may not include latent heat of phase change, depending on context), $J \cdot kg^{-1} \, K^{-1}$
c_{app}	Apparent specific heat including latent heat of phase change, $J \cdot kg^{-1} \, K^{-1}$
c_f	Specific heat of frozen food (not including latent heat), $J \cdot kg^{-1} \, K^{-1}$
c_u	Specific heat of unfrozen food, $J \cdot kg^{-1} \, K^{-1}$
C	Molar concentration, $mol \cdot m^{-3}$
$\mathbf{C}$	Capacitance matrix, $J \cdot K^{-1}$
D	Diffusivity, $m^2 \, s^{-1}$
D_v	Diffusivity of water vapour in air, $m^2 \, s^{-1}$
D_w	Effective diffusivity of moisture, $m^2 \, s^{-1}$
e_{ij}	Deviatoric strain
e_v	Fractional volumetric expansion due to freezing

e_x, e_y, e_z Fractional linear expansion due to freezing

E_c Cooling time shape factor (cooling time of infinite slab with same thickness/cooling time of object)

E_f Freezing time shape factor (freezing time of infinite slab with same thickness/freezing time of object)

E   Electric field, $V \cdot m^{-1}$

f   Forcing vector, W

F Frozen fraction

$f_T(H)$ Function to calculate temperature from enthalpy, K

G Shear modulus, Pa

h Heat transfer coefficient, $W \cdot m^{-2} \, K^{-1}$

H Enthalpy, $J \cdot kg^{-1}$

i Dissociation constant (van't Hoff factor); index; imaginary unit $\left(\sqrt{-1}\right)$

J Diffusion flux, $kgmole \cdot m^{-2} \, s^{-1}$

J   Jacobian matrix

k Thermal conductivity, $W \cdot m^{-1} \, K^{-1}$

k_{eff} Effective thermal conductivity, $W \cdot m^{-1} \, K^{-1}$

k_g Mass transfer coefficient, $kg \cdot s^{-1} \, m^{-2} \, Pa^{-1}$

K Bulk modulus, $\rho dP/d\rho$, Pa

K_R The ratio of MCP to minimum distance from surface to thermal centre

K_d Distribution coefficient during freezing of a solution (mass fraction of solids in ice/mass fraction of solids in liquid)

K_R Surface curvature ($K_R = 2/R$ for a sphere of radius R)

K   Conductance matrix, $W \cdot K^{-1}$

L Thickness, m

L_f Latent heat of freezing (positive) per unit mass of food, $J \cdot kg^{-1}$

m Mass, kg

$\dot{m}$ Mass flux, $kg \cdot s^{-1} \, m^{-2}$

M Molar mass, $kg \cdot kgmol^{-1}$

M   Mass matrix, kg

N Production rate, s^{-1}

n   Unit vector outward normal to surface, m

N   Shape function vector

P Pressure, Pa

P_a Water vapour pressure in the surrounding air, Pa

P_s Water vapour pressure at the food surface, Pa

P_{atm} Atmospheric pressure, Pa

P^* Saturation vapour pressure, Pa

P^*_{liq} Saturation vapour pressure of liquid water, Pa

P^*_{ice} Vapour pressure of ice, Pa

Pk Plank's number, $c_u(T_i - T_f)/\lambda_f$ or $c_u(T_i - T_f)/L_f$

q Heat flux, $W \cdot m^{-2}$

q   Heat flux vector, $W \cdot m^{-2}$

Q Heat load, W

Q_{cum} Cumulative heat load, J

r	Radial coordinate, m
r_f	Radius or half-dimension of the frozen zone, m
R	Radius or smallest half-dimension, m
R_g	Gas constant (8,314 J·K^{-1} kgmol^{-1} or Pa·m^3 K^{-1} kgmol^{-1})
R_m	Mean conducting path (MCP), m
S	Surface
S_q	Internally generated heat per unit volume, W·m^{-3}
$\mathbf{S}_v$	Momentum source vector, N·m^{-3}
Ste	Stefan number, $c_f(T_f - T_a)/L_f$
t	Time, s
t_f	Freezing time (time for the thermal centre temperature to fall from T_i to T_c), s
t_t	Thawing time, s
T	Temperature, K
T_0	Freezing point of pure water (273.15 K)
T_a	Temperature of the freezing or thawing medium, K
T_c	Final product centre temperature, K
T_e	Final average product temperature
T_f	(initial) Freezing point, K
T_{fm}	Nominal mean freezing temperature in Pham's (1986) freezing time method, K
$T_{g'}$	Glass transition temperature, K
T_i	Initial food temperature, K
T_m	Mean product temperature. K
T_s	Boundary (surface) temperature, K
$\mathbf{T}$	Nodal temperature vector, K
u	Radial displacement, m
$\mathbf{v}$	Velocity vector, m·s^{-1}
$v_f, \mathbf{v}_f$	Velocity of freezing front, m·s^{-1}
v_w	Molar volume of water, m^3(kgmol)$^{-1}$
V	Volume, m^3
w	Kirchhoff function, $w = \int_{T_{ref}}^{T} k d\theta$, W·m^{-1}
W	Mass ratio of total moisture to dry solids, kg·(kg dry solid)$^{-1}$
W_b	Mass ratio of bound moisture to dry solids, kg·(kg dry solid)$^{-1}$
W_l	Mass ratio of free liquid moisture to dry solids, kg·(kg dry solid)$^{-1}$
$W_{l, eq}$	Mass ratio of equilibrium free liquid moisture to dry solids, kg·(kg dry solid)$^{-1}$
W_{ice}	Mass ratio of ice to dry solids, kg·(kg dry solid)$^{-1}$
x	Mole fraction
x, y, z	Space coordinates, m
x_1, x_2, x_3	Space coordinates, m
x_f	Coordinate of freezing interface, m
X	Mass fraction
y	Mole fraction in the gas phase
Z	Geometry factor relating frozen fraction to freezing front position r_f

α	Thermal diffusivity $k/\rho c$, $\mathrm{m^2 s^{-1}}$
α	Time stepping parameter
β	Ratio of larger to smaller dimension for a two-dimensional shape
β_1	of larger to smaller dimension for a two-dimensional shape ($=\beta$), or ratio of second largest to smallest dimension for a three-dimensional shape
β_2	Ratio of largest to smallest dimension for a three-dimensional shape
β_T	Total thermal expansion coefficient, $\mathrm{K^{-1}}$
δ	Thickness of layer, or depth below surface, m
δ_{ij}	Kronecker delta (1 if $i=j$, otherwise 0)
ε_{ij}	ij-th component of strain
ε	Permittivity, $\mathrm{F \cdot m^{-1}}$
ε', ε''	Real and imaginary component of complex permittivity, $\mathrm{F \cdot m^{-1}}$
ϕ	Volume fraction
Φ	Phase field variable
ϕ_g	Void fraction
φ	Level set variable
λ_f	Latent heat of freezing (positive) per unit mass of water, $\mathrm{J \cdot kg^{-1}}$
λ_v	Latent heat of vaporisation (positive) per unit mass of moisture, $\mathrm{J \cdot kg^{-1}}$
λ_{vi}	Latent heat of sublimation (positive) from ice per unit mass of moisture, $\mathrm{J \cdot kg^{-1}}$
λ_{vl}	Latent heat of evaporation (positive) from liquid water per unit mass of moisture, $\mathrm{J \cdot kg^{-1}}$
μ	Viscosity, $\mathrm{kg \cdot m^{-1} \, s^{-1}}$
ρ	Density, $\mathrm{kg \cdot m^{-3}}$
Ω	Space domain
σ_{ij}	Stress, Pa
σ_{sl}	Surface energy, $\mathrm{J \cdot m^{-2}}$
θ	$T - T_0$, Temperature in °C (for subscripts see definitions of T)
τ	Tortuosity factor
ω	Angular frequency, $\mathrm{rad \cdot s^{-1}}$

Subscripts

0	At 0 °C
a	Air or surroundings
ash	Ash
av	Average
b	Bound water
$carb$	Carbohydrates
ds	Dry solid
eff	Effective
f	Frozen food, frozen part of the food
fat	Fat

fib	Fibre
i	Nodal index
ice	Ice
l	Free (unbound) liquid moisture
liq	Liquid phase
M–E	Maxwell–Eucken model
o	Original
par	Parallel model of thermal conductivity
ser	Series model of thermal conductivity
prot	Protein
r	Radial component
ref	Reference
s	Surface
sol	Solute
t	Tangential component
u	Unfrozen food
uw	Liquid (unfrozen) water
v	Water vapour
w	Water; total moisture (liquid, ice and water vapour)

Superscripts

el	Element
m	Iteration counter
(m)	Mechanical strain
New	New value (at end of next time interval)
(T)	Thermal strain

Chapter 1
Introduction to the Freezing Process

The focus of this work is on the freezing of solid or liquid foods which are in a stationary state. The freezing of liquid foods in motion is considered briefly in Chap. 6.

The starting point in any practical calculation is to understand thoroughly the physical situation that we are modelling, so that we can identify the influential factors. Food freezing is a complex process that may involve several physical phenomena and be influenced by many factors. We shall, therefore, start by considering a much simpler situation, and then gradually incorporate other factors.

Consider a small droplet of water suspended in a cold, sub-zero environment. The droplet is so small that its temperature can be assumed to be uniform at all times. The only impediment to instant freezing is the finite *external (surface) resistance* to heat transfer, which controls the rate at which heat can flow from the droplet's surface to the air. This surface resistance is due to the finite thermal conductivity of the air around the droplet. The droplet will generally start at some temperature higher than the freezing point, so for some time it will cool down without any freezing, a process known as *sensible cooling* because the temperature change can be sensed or measured with a thermometer. Even when the freezing point is reached, it will usually continue to cool without phase change—this is called *supercooling*. Only when several degrees of supercooling have been reached will *nucleation* and phase change occur. Because freezing releases a large amount of latent heat which has to be transferred to the surroundings, the temperature of the droplet will rise to the freezing point T_f and stay there for some time, during which ice and liquid water co-exist. Once phase change is completed, temperature will fall again until the ice is equilibrated with the external environment. The droplet thus goes through three stages: *precooling*, phase change and *subcooling* (Fig. 1.1).

Next we consider a larger mass of still (nonflowing, unstirred) water being frozen in cold air, such as a cup of water. The main difference between this case and the droplet we considered earlier is that because of the size of the product, a significant temperature profile occurs, with the surface being colder than the inner regions. In other words, *internal resistance* to heat transfer becomes an important factor. After some sensible cooling, including supercooling, nucleation will happen at the surface and the temperature near the surface jumps quickly to the freezing point. A layer of ice forms at the surface and gradually thickens. The interface between the frozen

Q. T. Pham, *Food Freezing and Thawing Calculations*,
SpringerBriefs in Food, Health, and Nutrition, DOI 10.1007/978-1-4939-0557-7_1,
© The Author 2014

Fig 1.1 Stages in the freezing
of a small water droplet

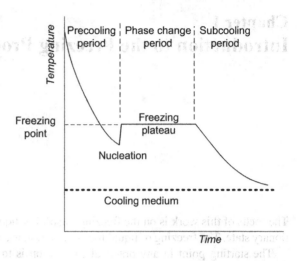

and unfrozen regions is called the *freezing front* and its temperature is the freezing temperature. The frozen region is below the freezing point while the inner, unfrozen region is above it. Both regions continue to cool as heat is continually conducted from the inside towards the surface. The freezing front gradually moves deeper into the product, usually by the growth of the crystals that formed near the surface. Unless freezing is extremely rapid or there are barriers to crystal growth such as plastic films or air gaps, once surface nucleation has occurred crystal growth will become the principal or only mechanism, and no further supercooling or nucleation will occur because phase change no longer relies on nucleation.

Once the freezing front reaches the centre, phase change is completed. Thereafter, some further sensible cooling will proceed until the ice is equilibrated with the external environment. Each point in the product has gone through the three stages of precooling, phase change and subcooling, at different times depending on the distance from the surface (Fig. 1.2). The surface crosses the freezing temperature fairly quickly, while the centre cools down to a near-freezing temperature, then remains in that vicinity, in an unfrozen state, for a certain duration. The centre temperature still displays an apparent 'freezing plateau' similar to the droplet's, but its physical meaning is not identical, since, in this case the plateau starts some time after nucleation occurs at the surface and before phase change commences at the centre. It simply means that, because the unfrozen core is surrounded by a freezing front whose temperature is the freezing point, any point in it (including the centre) will cool down and approach that temperature but cannot cool any further until the freezing front has crossed it. For the product as a whole, there are no well-defined precooling, phase change and subcooling periods. Nevertheless, we will see that these concepts will be adapted for the calculation of freezing time in solid foods.

If the product being frozen is food instead of pure water, things get still more complicated because the moisture in the food contains solutes such as salts or sugars. The freezing point is accordingly depressed. As ice forms and separates out,

Fig. 1.2 Stages in the freezing of still water (*solid line*) and that of a solid food (*dashed line*)

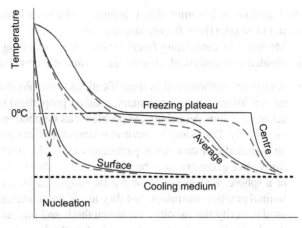

the remaining solution becomes more concentrated and its equilibrium temperature (freezing point) decreases further. As a result ice continues to form as the local temperature falls below the initial freezing point. There can still be still a freezing plateau at the centre because the (continually shrinking) unfrozen core cannot cool below the initial freezing point, but at the end of this plateau the centre temperature falls gradually instead of sharply (Fig. 1.2). The freezing front is not sharp but diffuse.

In industry, the freezing time is usually defined as the time for the thermal centre (the slowest cooling point) of the food to reach a specified temperature, say $-10\,^{\circ}\text{C}$ or more often $-18\,^{\circ}\text{C}$. The location of the thermal centre is not always obvious. For symmetrical products subjected to symmetrical or uniform surface conditions, it coincides with the geometric centre. However, for irregular shaped products the thermal centre is not easy to locate and may also depend on operating conditions. A very complex shape such as an animal carcass may have several local thermal centres, for example one in the leg and one in the shoulder. In such cases experimentation or numerical simulation are necessary to determine the thermal centres.

Although supercooling can be quite pronounced, most food calculation methods ignore its effect on the freezing time. The freezing process is assumed to be governed principally by heat transfer. Numerical calculations have shown that this is generally a valid assumption (Pham 1989) but there are exceptions.

Phase change is usually not the only physical phenomenon that occurs in food freezing. It is often accompanied by movement of water and solutes from the macro scale through the cellular scale (exchange of water and solutes between the cells and the extracellular space) to the molecular scale (e.g. movement of water molecules towards the growing crystal surface or solutes away from it). Application of high pressure will change the equilibrium properties of the food. In addition to ice formation there may be other physical transitions such as glass transition, where the liquid phase becomes almost infinitely viscous without crystallising. A precise model must take these factors into account.

Methods for calculating freezing times and modelling the freezing process can be divided into analytical, approximate, empirical and numerical:

- Analytical methods makes simplifications to the physical situation (environmental conditions, product geometry, product properties) until a sufficiently simple set of equations describing the problem can be obtained and solved analytically, i.e. exactly. For example, environmental conditions and product geometry may be assumed constant, some parameters may take limiting values of zero or infinity, and geometry may be assumed to be that of a slab, an infinite cylinder or a sphere. Because of these drastic simplifications, analytical methods are of limited practical usefulness but they are of great interest to researchers. They are used to verify the validity of other methods and may serve as a starting point for developing approximate and empirical methods.
- Approximate methods avoid excessive simplifications but still rely on some approximate models of reality. Usually, they divide the freezing process into distinct sensible cooling and phase change periods, and assume that the food's physical properties are constant during each of these periods. In practice, sensible cooling and phase change periods overlap and properties vary continuously with temperature. Hopefully by making judicious approximations the resulting predictions are still accurate enough to be of practical use.
- Empirical methods may or may not be based partly on theory (the former is usually the case), but include parameters that are derived from curve-fitting data obtained by experimentation or numerical calculations. There is no limit to the number of factors that can be taken into account or their ranges, as long as enough empirical parameters are used. However, these equations tend to perform poorly when conditions are outside those already tested.
- Numerical methods involve the discretisation of space and time into small elements or control volumes, and then rigorously solving the differential equations describing the evolution of temperature and other field variables in these partitions. They can be carried out to any of precision required as long as sufficient computational resources are available, and take into account any number of physical effects as long as their mechanism and governing mathematics are understood. Practical engineers will be more interested in approximate, empirical or numerical methods, while researchers find analytical methods useful as a starting point for deriving approximate or empirical methods, for estimating the effect of varying design and operating variables, and for verifying the theoretical soundness of approximate, empirical and numerical methods. In particular, to be valid the latter must agree with analytical methods in known limiting cases.

'Garbage in, garbage out' is a common saying in computer science and it applies equally to food freezing calculations. Even with the most sophisticated methods (numerical calculations) the predictions are reliable only if the inputs are correct. In this case the 'inputs' are chiefly the physical properties of the product. We shall therefore begin by addressing this topic in the next chapter.

Chapter 2
Heat Transfer Coefficient and Physical Properties

2.1 Introduction

Before freezing time and heat load can be calculated, the physical parameters or inputs must be available. These can be classified into two groups: environmental factors and food properties. The former comprise the environment temperature and the heat transfer coefficient. The latter is often difficult to determine and its calculation is a vast subject, but only a brief treatment will be given in this work.

The relevant physical properties of the food are the freezing point, enthalpy, specific heat, density and thermal conductivity. The last four properties are functions of temperature and there is usually a discontinuity or large change around the freezing point. If data are available, we still have to put them in a usable form by using some curve-fitting (regression) equation. If reliable data on the properties are not available, we have to estimate them from composition data. This chapter describes some of the ways this can be done. The property prediction methods described in this chapter are certainly not the only ones and not necessarily the most advanced available, and there will be methods more suited to specific types of foods, but they are conceptually simple and will in most cases provide satisfactory answers. A more comprehensive review may be obtained from Gulati and Datta (2013).

2.2 Heat Transfer Coefficient

The heat transfer coefficient h is the ratio of heat flux q (heat flow per unit area) to the difference between the temperature T_s of the surface and that of the cooling medium, T_a:

$$h = \frac{q}{T_s - T_a} \qquad (2.1)$$

Q. T. Pham, *Food Freezing and Thawing Calculations*,
SpringerBriefs in Food, Health, and Nutrition, DOI 10.1007/978-1-4939-0557-7_2,
© The Author 2014

Its inverse $1/h$ is termed the surface resistance to heat transfer. For an unwrapped product in air, h includes contributions from convection, radiation and evaporative cooling (see e.g. Davey and Pham 1997) and will be denoted by the symbol h_{surf}:

$$h_{surf} = h_{conv} + h_{rad} + h_{evap} \qquad (2.2)$$

The inverse of h_{surf} measures the resistance to heat transfer at the food–fluid interface. The convective contribution depends on the geometry of the product, the properties of the surrounding fluid, the flow pattern and the degree of turbulence. For many frequently occurring configurations the convective heat transfer coefficient can be calculated from empirical correlations. It can also be calculated by computational fluid dynamics, provided a suitable turbulence model is used (Hu and Sun 2001; Pham et al. 2009). The radiative contribution depends on the temperature and emissivity of the product surface and those of the radiation source or sink (such as the surrounding walls). The evaporative cooling contribution depends on the temperature and humidity at the product surface and those of the surroundings. In immersion freezing the radiation and evaporative cooling contributions can be neglected.

If the product is wrapped, heat from the product must be released by passing it through layers of trapped air (if any), wrapping materials and the boundary layer at the fluid–solid interface; therefore the effective heat transfer coefficient will be given by

$$h = \left(\frac{\delta_{air}}{k_{air}} + \frac{\delta_{wrap}}{k_{wrap}} + \frac{1}{h_{surf}} \right)^{-1} \qquad (2.3)$$

where δ is the thickness of the trapped air or wrapping.

The heat transfer coefficient is frequently the most uncertain parameter in a freezing calculation. However, errors in h are not serious if the Biot number is large, which indicates that internal resistance to heat transfer rather than h is the controlling factor (more of this in Sect. 3.2.3). For most of the non-numerical methods covered in this work, h is assumed to be constant throughout the cooling and freezing process, although in practice it may vary.

The heat transfer coefficient is highly dependent on process conditions and its estimation is a vast and complex subject. The reader is referred to Kondjoyan (2006) for a review in the food refrigeration field. Becker and Fricke (2004) presented some experimental values. Typical ranges are shown in Table 2.1.

2.3 Density

In one unit mass of a composite material, each component i contributes mass X_i and volume X_i/ρ_i. The density of the material is therefore the total mass (1) divided by the total volume:

$$\rho = \frac{1}{\sum_i \dfrac{X_i}{\rho_i}} \qquad (2.4)$$

Table 2.1 Typical heat transfer coefficients for common equipments. (Source: Karel and Lund 2003; Sinha 2011; Pham unpublished)

Type of equipment	Heat transfer coefficients ($W \cdot m^{-2}K^{-1}$)
Still air chillers and freezers	5–10
Air blast chillers and freezers	15–30
Impingement chillers and freezers	50–200
Fluidized bed chillers and freezers	50–250
Immersion chillers and freezers	100–300
Plate chillers and freezers	500–1000

Fig. 2.1 Density of liquid water at atmospheric pressure: Kell (Eq. 2.5), Choi & Okos's equation and data from Hare and Sorensen (1987), Sotani et al. (2000) and Lide (2009)

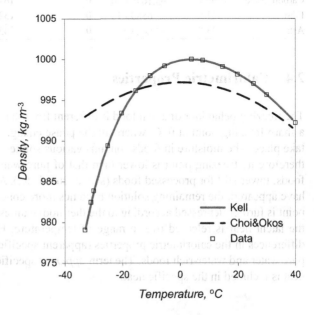

Polynomial expressions for the density of food components as functions of temperature were presented by Choi and Okos (1986) and the coefficients are shown in Table 2.2. For water, Choi & Okos's coefficients are given only for completeness, as the equation is not very accurate. Values for the density of water are available in the CRC Handbook (Lide 2009), on the NIST website (2013a) and in Wagner and Pruss (2002). An accurate regression equation for water density at atmospheric pressure down to −30 °C was given by Kell (1975) (Fig. 2.1):

$$\rho_{uw} = \frac{a_0 + a_1\theta + a_2\theta^2 + a_3\theta^3 + a_4\theta^4 + a_5\theta^5}{1+b\theta} \qquad (2.5)$$

where $a_0 = 999.83952$, $a_1 = 16.945176$, $a_2 = -7.9870401 \times 10^{-3}$, $a_3 = -4.6170461 \times 10^{-5}$, $a_4 = 1.0556302 \times 10^{-7}$, $a_5 = -2.8054253 \times 10^{-10}$, $b = 0.016879850$.

Table 2.2 Coefficients of the equation $y = A + B\theta + C\theta^2$ for the dependence of density on temperature (Choi and Okos 1986) and predicted values at −40, −20 and 20 °C. Predictions from Eq. 2.5 (recommended) for water are also included

Component	A	B	C	Value at −40 °C	Value at −20 °C	Value at 20 °C
Water	997.18	3.1439×10^{-3}	-3.7574×10^{-3}	(991)	(996)	(996)
Water (Eq. 2.5)	–	–	–	962	994	998
Ice	916.89	-1.3071×10^{-1}	0	922	920	–
Protein	1329.9	-5.1840×10^{-1}	0	1351	1340	1320
Fat	925.59	-4.1757×10^{-1}	0	942	934	917
Carbohydrate	1599.1	-3.1046×10^{-1}	0	1612	1605	1593
Fibre	1311.5	-3.6589×10^{-1}	0	1326	1319	1304
Ash	2423.8	-2.8063×10^{-1}	0	2435	2429	2418

2.4   Calorimetric Properties

The freezing behaviour of a real food is different from that of water. Pure water has a sharp freezing point at 0 °C, where all the phase changes and latent heat releases take place. The moisture in foods contains various solutes such as salts and sugars; therefore its freezing point is lower than that of pure water, about −1 °C for fresh foods, lower still for processed foods (salted, cured etc.). After the first ice crystals have appeared, the remaining solution becomes more concentrated and its freezing point is further depressed according to the thermodynamics of solutions. Therefore, the latent heat is released over a range of temperature. Figure 2.2 illustrates the differences in the calorimetric properties (apparent specific heat and latent heat) of pure water and water-rich foods. The term *apparent* specific heat implies that latent heat is included in the specific heat.

2.4.1   Freezing Point

Because phase change in food freezing takes place over a range of temperature, we also use the term *initial freezing point* to refer to the temperature where the ice crystal first appears (in the absence of supercooling). It can be calculated from the concentration of solute using the thermodynamic relationship

$$-\ln a_w = \frac{M_w \lambda_f}{R_g}\left(\frac{1}{T_f} - \frac{1}{T_0}\right) \tag{2.6}$$

where a_w is the water activity, M_w the molar mass of water (18.02 kg·kmol^{-1}), λ_f the latent heat of freezing per unit mass of water (334 kJ·kg^{-1}), R_g the gas constant (8.314 kJ·K^{-1}kmol^{-1}), T_f the freezing point of the solution and T_0 that of pure water

Fig. 2.2 Apparent specific
heat of water–ice and foods

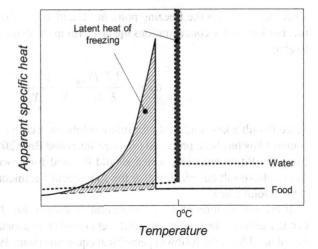

(273.15 K). For dilute solutions, Raoult's law applies, and water activity can be
calculated from

$$a_w = 1 - i x_{sol} \qquad (2.7)$$

where x_{sol} is the solute mole fraction in the solution and i the van't Hoff factor,
which accounts for molar or ionic dissociation in the solution. The solute mole fraction x_{sol} can be calculated as follows. Let X_s be the mass fraction of total solids, X_w
that of water and X_{sol} that of solutes. In the widely used bound water model, some
of the water is bound to the solid phase and does not participate in the solvation or
freezing process (Schwartzberg 1976), while the rest is *free water*. Let the mass
fraction of bound water (as a fraction of total food mass) be X_b, then that of free
water will be $X_w - X_b$. The solute mole fraction in the free water will therefore be

$$x_{sol} = \frac{\dfrac{X_{sol}}{M_{sol}}}{\dfrac{X_{sol}}{M_{sol}} + \dfrac{X_w - X_b}{M_w}} \qquad (2.8)$$

Schwartzberg et al. (2007) showed that Eq. 2.6 can be well approximated by

$$\theta_f \approx -\frac{R_g T_0^2}{M_w \lambda_f} \frac{\dfrac{M_w}{M_{sol}} i X_{sol}}{X_w - X_b + \dfrac{1}{2} \dfrac{M_w}{M_{sol}} i X_{sol}} \qquad (2.9)$$

where $\theta_f \equiv T_f - T_0$ is the freezing point in °C and M_{sol} is the molecular mass of solute. For low solute concentrations the last term in the denominator can be neglected to give

$$\theta_f \approx -\frac{R_g T_0^2 i X_{sol}}{\lambda_f M_{sol}} \frac{1}{X_w - X_b} \qquad (2.10)$$

Since Raoult's law holds only for dilute solutions, a correction for deviations from Raoult's law has been proposed (van der Sman and Boer 2005; van der Sman 2008) using the Pitzer equation, which should be used for processed meats which frequently have salt added. However the subsequent treatment in this chapter will assume Raoult's law.

If the composition of a food is accurately known, Eq. 2.9 or 2.10 will give accurate freezing time predictions (Miles et al. 1997). Reasonable predictions can also be obtained from the following empirical equation (Pham 1996), based on available data for meat, fish, fruit and non-fat dairy foods:

$$\theta_f \approx -\frac{1}{X_w}(4.66 X_{other} + 46.4 X_{ash}) \qquad (2.11)$$

where X_{ash} is the ash (mineral) mass fraction in the food and X_{other} is the mass fraction of components (such as carbohydrates) other than ash, water, fat or proteins.

2.4.2 Bound Water

Bound water may be calculated from

$$X_b = b X_{ds} \qquad (2.12)$$

where X_{ds} is the total dry solids mass fraction and b is the bound water to solids mass ratio. According to Duckworth (1971) b ranges from $0.24 - 0.27$ for meat and fish muscle. Pham (1987a) found the following mean (minimum, maximum) values for b: for meat and fish, 0.217 (0.143, 0.318); for egg, 0.106 (0.103, 0.109); for bread, 0.127 (0.111, 0.143); for Tylose (a methyl cellulose gel often used as a meat analogue in freezing experiments), 0.402 (0.352, 0.430). Fikiin (1998) reported b values of 0.257 for meat, $0.270 - 0.280$ for fish, 0.275 for egg white, 0.167 for yeast, 0.080 for green peas, 0.117 for spinach.

More accurate predictions may be obtained by taking into account the composition of the food solids. Pham (1987a) found that for meat the bound water/protein ratio is about 0.4. Miles et al. (1997) and van der Sman and Boer (2005) proposed

$$X_b = 0.3 X_{prot} + 0.1 X_{carb} \qquad (2.13)$$

Pham (1996) proposed

$$\frac{X_b}{X_w} = 0.342(1 - X_w) - 4.51X_{ash} + 0.167X_{prot} \tag{2.14}$$

The last equation has the advantage of correctly predicting that $X_b \to 0$ when X_w tends to 0 or unity (pure water).

2.4.3 Frozen Fraction

As we have seen earlier, after the food passes through its initial freezing point, ice continues to form as temperature falls, leaving behind a more and more concentrated solution. This solution is in thermodynamic equilibrium with the ice and therefore its freezing point will be T or θ. Thus the solution's concentration and its activity are related to temperature by the same equations that determine initial freezing point (Eq. 2.6, 2.9 or 2.10), with T_f or θ_f replaced by T or θ and X_w replaced by $X_w - X_{ice}$, where X_{ice} is the mass fraction of ice. For example, Eq. 2.10 becomes:

$$\theta \approx -\frac{R_g T_0^2 i X_{sol}}{\lambda_f M_{sol}} \frac{1}{X_w - X_{ice} - X_b} \tag{2.15}$$

Then dividing Eq. 2.10 by Eq. 2.15 and re-arranging gives

$$X_{ice} \approx (X_w - X_b)\left(1 - \frac{\theta_f}{\theta}\right) \tag{2.16}$$

Although this equation is accurate only at temperatures not far below the freezing point, it is often used to estimate the ice fraction and hence enthalpy due to its simplicity. For processed foods with salts added, van der Sman's approach (van der Sman and Boer 2005; van der Sman 2008), which corrects for deviations from Raoult's law, can be used.

Others have proposed empirical equations derived from fitting experimental data. Fikiin (1998) recommended the following equation by Tchigeov (1956, 1979) for meat, fish, milk, eggs, fruits and vegetables, which requires only the freezing point and total moisture content and can thus be used when full composition is not known:

$$X_{ice} = \frac{1.105}{1 + \dfrac{0.7138}{\ln(\theta_f - \theta + 1)}} X_w \tag{2.17}$$

Fig. 2.3 Frozen fraction for two foods with freezing points differing by 0.5 K

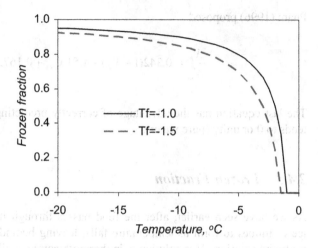

Fig. 2.4 Latent heat release rate for two foods with freezing points differing by 0.5 K

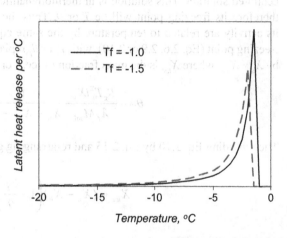

The equation is reported to be satisfactory for $-45\,°C \leq \theta \leq \theta_f$ and $-2\,°C \leq \theta_f \leq -0.4\,°C$.

Errors due to uncertainty in freezing point Often the freezing point T_f or θ_f is only known to within ±0.5 K, so there is a large relative uncertainty in the last term in Eq. 2.16 and hence in the ice content and resulting enthalpy and specific heat curves (Figs. 2.3 and 2.4). It is possible to estimate the effect of this uncertainty on freezing time and heat load calculations. If freezing is stopped at a final product temperature $\theta_e = -18\,°C$, the total latent heat load will be proportional to the amount of ice formed at θ_e, i.e. proportional to $1 - \theta_f/\theta_e$. If the uncertainty in θ_f is $\delta\theta_f$, the relative error in the total latent heat load will be $(\delta\theta_f/\theta_e)/(1 - \theta_f/\theta_e)$. For $\theta_f = -1.0\,°C$, $\delta\theta_f = 0.5$ K, $\theta_e = -18\,°C$, the latent heat error will be around 3 %. The error in the total heat load (including sensible heat) will be somewhat smaller.

Again ignoring sensible heat, the instantaneous rate of heat removal is inversely proportional to the temperature difference between the food and its surroundings,

$\theta - \theta_a$, therefore the freezing time is proportional to $\int\limits_{\theta_f}^{\theta_e} \dfrac{dX_{ice}}{\theta - \theta_a}$, i.e. to $\int\limits_{\theta_f}^{\theta_e} \dfrac{d(1 - \theta_f/\theta)}{\theta - \theta_a}$

or $\int\limits_{\theta_f}^{\theta_e} \theta_f \dfrac{d(1/\theta)}{\theta - \theta_a}$ from Eq. 2.16. Integration shows that for $\theta_f = -1.0\,°C$, $\delta\theta_f = 0.5$ K,

$\theta_e = -18\,°C$, $\theta_a = -30\,°C$, the freezing time error will be around 1.3 %. If sensible heat is taken into account the error will be smaller still. The lowering of the freezing point tends to increase freezing time, but this is compensated by a lesser latent heat load, therefore the freezing time error is small. In conclusion we can say that errors in the initial freezing point are likely to have only a small effect on freezing time.

2.4.4   Enthalpy

Enthalpy H ($J\cdot kg^{-1}$) is the heat content per unit mass of food at a given temperature. It is measured from an arbitrary reference temperature, since in all calculations we are concerned only with changes in enthalpy. In freezing calculations the usual reference temperature is $T_{ref} = -40\,°C$.

Once the ice content has been calculated as shown in the previous section, the food's enthalpy can be obtained simply by adding those of the components:

$$H = X_{uw} H_{uw} + X_{ice} H_{ice} + X_{prot} H_{prot} + X_{fat} H_{fat}$$
$$+ X_{carb} H_{carb} + X_{fib} H_{fib} + X_{ash} H_{ash} \qquad (2.18)$$

where for each component i apart from water:

$$H_i = \int\limits_{T_{ref}}^{T} c_i dT \qquad (2.19)$$

where c_i is the specific heat ($J\cdot kg^{-1}K^{-1}$) of component i. Equation 2.19 assumes that the mixture is ideal, i.e. that there is no heat of mixing effect. The enthalpy of unfrozen water H_{uw} is calculated differently from that of the other components, because water exists in two phases and we have taken ice at T_{ref} to be the reference state. Since liquid water at temperature T can be obtained by taking ice at the reference temperature, heating it to T_0 ($0\,°C$), melting it, then bringing the liquid water to temperature T, its enthalpy will be given by:

$$H_{uw} = \int\limits_{T_{ref}}^{T_0} c_{ice} dT + \lambda_{f0} + \int\limits_{T_0}^{T} c_{uw} dT \qquad (2.20)$$

Fig. 2.5 Specific heat of water according to Eq. 2.21, Choi and Okos's equation and averaged experimental data (Wagner and Pruss 2002; Holten et al. 2013).

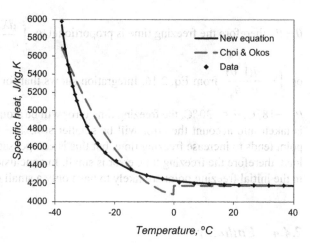

where λ_{f0} is the latent heat of freezing at $0\,°C$. Values of specific heats for water, ice, proteins, fat, carbohydrates, fibre and ash are listed by Choi and Okos (1986) as second-order polynomial functions of T, which can be readily integrated to yield enthalpy values according to the above equations. The coefficients of the functions are given in Table 2.3 together with typical values. For liquid water there is a step jump at $0°$ and the equation does not agree well with accepted values (Fig. 2.5), therefore the following equation should be used instead:

$$c_{uw} = 6.6353 \times 10^5 x^2 - 1.2132 \times 10^4 x + 4231.0 \qquad (2.21)$$

where $x = 1/(\theta + 54.15)$. This agrees with accepted values (Wagner and Pruss 2002; Holten et al. 2013) for $-40\,°C \le \theta < 100\,°C$ to within $1\,\%$.

Over a limited temperature range the specific heats of the components are often assumed to be constant, and if furthermore Raoult's law holds (solution is ideal) the following approximate equations are obtained (Schwartzberg 1976; Pham 1987a):

$$H(T \le T_f) = A + c_f \theta - \frac{B}{\theta} \qquad (2.22)$$

$$H(T > T_f) = H_{T=T_f} + c_u(\theta - \theta_f) \qquad (2.23)$$

where A is an arbitrary constant and

$$B = -(X_w - X_b)\lambda_{f0}\theta_f \qquad (2.24)$$

Table 2.3 Coefficients of the equation $y = A + B\theta + C\theta^2$ for the dependence of specific heat on temperature (Choi and Okos 1986) and predicted values at −40, −20 and 20 °C. Predictions from Eq. 2.21 (recommended) for water are also included

Component	A	B	C	Value at −40 °C	Value at −20 °C	Value at 20 °C
Water (−40 to 0 °C)	4081.7	−5.3062	9.9516×10^{-1}	5886	4586	–
Water (0 to 150 °C)	4176.2	-9.0864×10^{-2}	5.4731×10^{-3}	–	–	4177
Water (Eq. 2.21)	–	–	–	6688	4445	4188
Ice	2062.3	6.0769	0	1819	1941	–
Protein	2008.2	1.2089	-1.3129×10^{-3}	1958	1983	2032
Fat	1984.2	1.4733	-4.8008×10^{-3}	1918	1953	2012
Carbohydrate	1548.8	1.9625	-5.9399×10^{-3}	1461	1507	1586
Fibre	1845.9	1.8306	-4.6509×10^{-3}	1765	1807	1881
Ash	1092.6	1.8896	-3.6817×10^{-3}	1011	1053	1129

The parameters of Eq. 2.22 and 2.23 can be derived from composition (Pham 1996) or by curve-fitting experimental data. Pham (1987a) tabulates values of the parameters A, B, θ_f and c_f for several types of foods.

Due to the limitations of Raoult's law, the above enthalpy equations may not always agree well with experimental data for temperatures below freezing. Furthermore, although the moisture content can easily be measured, full composition is not always available. Fikiin and Fikiin (1999) proposed the following empirical equation which only requires the freezing point and total moisture content to be known:

$$H(T \leq T_f) = -C(T_f - T)^D \tag{2.25}$$

where $C = 225.25 X_w - 13.105, \quad D = \dfrac{0.046}{X_w} + 0.122.$

2.4.5 Sensible Specific Heat

Some analytical and approximate methods for calculating freezing/thawing times and heat loads require values of the sensible or "true" frozen and unfrozen specific heats c_f and c_u (excluding latent heat of phase change) at certain temperatures. These can be calculated from

$$c_f = X_{uw}c_{uw} + X_{ice}c_{ice} + X_{prot}c_{prot} + X_{fat}c_{fat} + X_{carb}c_{carb} + X_{fib}c_{fib} + X_{ash}c_{ash} \tag{2.26}$$

$$c_u = X_{uw}c_{uw} + X_{prot}c_{prot} + X_{fat}c_{fat} + X_{carb}c_{carb} + X_{fib}c_{fib} + X_{ash}c_{ash} \tag{2.27}$$

where X_{uw} and X_{ice} can be calculated from the equations in Sect. 2.4.3 and the components' c-values from those in Sect. 2.4.4.

2.4.6   Apparent Specific Heat

The derivative of enthalpy, $c_{app} \equiv dH/dT$, is termed the apparent (or effective) specific heat of the food, because it includes latent heat effects. The most general way to compute it is by numerical differentiation of the enthalpy–temperature function:

$$c_{app} = \frac{H(T+\delta T) - H(T-\delta T)}{2\delta T} \tag{2.28}$$

where δT is a small temperature interval. Numerical differentiation is difficult to carry out for pure water or dilute solutions at or near the freezing point, because of the abrupt changes in enthalpy. An arbitrary smearing out of the latent heat peak (freezing temperature range) will facilitate the process.

If it is desired to calculate apparent specific heats from composition, for temperatures above freezing $c_{app} = c_u$ as given in Eq. 2.27. Below the freezing point things get more complicated because of phase change. The contribution of water/ice to the apparent specific heat is

$$\frac{d}{dT}(X_{uw}H_{uw} + X_{ice}H_{ice}) = X_{uw}c_{uw} + X_{ice}c_{ice} + (H_{uw} - H_{ice})\frac{dX_{uw}}{dT} \tag{2.29}$$

since $dX_{ice} = -dX_{uw}$. From Eq. 2.20,

$$H_{uw} - H_{ice} = \lambda_{f0} + \int_{T_0}^{T}(c_{uw} - c_{ice})dT \tag{2.30}$$

while dX_{uw}/dT can be found, say, by differentiating Eq. 2.16:

$$\frac{dX_{uw}}{dT} = -\frac{dX_{ice}}{dT} \approx (X_w - X_b)\frac{T_0 - T_f}{(T_0 - T)^2} = (X_w - X_b)\frac{-\theta_f}{\theta^2} \tag{2.31}$$

Putting these expressions together, we obtain for the apparent specific heat below the freezing point:

$$c_{app}(T \leq T_f) = X_{uw}c_{uw} + X_{ice}c_{ice} + X_{prot}c_{prot} + X_{fat}c_{fat}$$

$$+ X_{carb}c_{carb} + X_{fib}c_{fib} + X_{ash}c_{ash} + \left(\lambda_{f0} + \int_{T_0}^{T}(c_{uw} - c_{ice})dT\right)(X_w - X_b)\frac{-\theta_f}{\theta^2}$$

$$\tag{2.32}$$

The following simple approximate expressions for c_{app} can be obtained by differentiating Eq. 2.22 and 2.23:

$$c_{app}(T \le T_f) = c_f + \frac{B}{\theta^2} \qquad (2.33)$$

$$c_{app}(T > T_f) = c_u \qquad (2.34)$$

Note: In the rest of this work the symbol c may be used to refer to the sensible specific heat (as defined in, say, Eq. 2.26, i.e. excluding latent heat of phase change) or to the apparent specific heat c_{app}, depending on context. For examples, in the heat conduction equation (Eq. 3.5) c will mean the sensible specific heat (c_u or c_f) if latent heat is treated separately as a heat source, or the apparent specific heat if the heat source term in Eq. 3.5 is dropped. The same goes for all numerical methods used to solve the heat conduction equation. When the apparent specific heat must be used (such as in the Apparent Specific Heat Method of Sect. 5.4.1.3), the symbol c_{app} will be employed. Above the freezing point, of course, $c = c_u = c_{app}$.

2.4.7 Calorimetric Properties at High Pressure

When required for the modelling of high pressure freezing and thawing, the calorimetric properties of a food can be calculated from those at atmospheric pressure by standard thermodynamic relationships.

2.4.7.1 Freezing Point at High Pressure

The freezing point depression due to pressure is given by the Clausius–Clapeyron equation

$$\frac{dP}{dT_f} = \frac{H_{ice} - H_{uw}}{T(V_{ice} - V_{uw})} = \frac{-\lambda_f}{\dfrac{1}{\rho_{ice}} - \dfrac{1}{\rho_{uw}}} \qquad (2.35)$$

Density difference and latent heat change markedly at high pressure (>10 MPa), therefore it is easier to rely on tabulated data for T_f vs. P (e.g. Lide 2009) or using the P–T relationship for the melting curve of ice (Wagner et al. 2011). The following regression equation is probably sufficiently close to Wagner's relationship (± 0.001 K) for food freezing studies:

$$\theta_f = 1.8287 \times 10^{-8} z^3 - 1.5252 \times 10^{-4} z^2 - 7.4445 \times 10^{-2} z + 0.010641 \quad (2.36)$$

where θ_f is the negative of the freezing point depression and $z = 10^{-6} P$ is the pressure in MPa.

2.4.7.2   Enthalpy–Temperature Curve

The rate of change of enthalpy with pressure in an isothermal process is given by the thermodynamic relationship

$$\left(\frac{\partial H}{\partial P}\right)_T = V - T\left(\frac{\partial V}{\partial T}\right)_P \qquad (2.37)$$

where $V = 1/\rho$ is the specific volume. Putting $(\partial V/\partial T)_P = V\beta_T$ where β_T is the thermal expansion coefficient and integrating this equation gives the enthalpy difference between a system at (T, P_1) and (T, P_2)

$$H(T, P_2) - H(T, P_1) = \int_{P_1}^{P_2} \frac{1 - T\beta_T}{\rho} \, dP \approx \frac{1 - T\beta_T}{\rho_{av}} (P_2 - P_1) \qquad (2.38)$$

where ρ_{av} is the average density between the two pressures. This equation enables the enthalpy–temperature curve at high pressure to be generated from that at atmospheric pressure. β_T can be calculated from the density–temperature relationships of the food's components (Eq. 2.5). This equation is useful only if there is no phase change during compression, i.e. if $T \geq T_{f, PI}$, because if there is phase change β_T is large and varies rapidly.

To calculate ρ_{av} in Eq. 2.38 we require values of density at the lower and higher pressure. For water this can be obtained from Holten et al. (2013). For ice, density at high pressure can be calculated from the bulk modulus, which is reported as (8.3 to 11.3)$\times 10^3$ MPa at $-5\,^{\circ}$C (Gold 1958) and (8.65$\pm$0.2)$\times 10^3$ MPa at $-16\,^{\circ}$C (Petrenko and Whitworth 1999).

2.5   Thermal Conductivity

Polynomial expressions for the dependence of the thermal conductivities of food components on temperature were presented by Choi and Okos (1986) and the coefficients are shown in Table 2.4. Predicting thermal conductivity k from composition is more difficult than predicting enthalpy or specific heat, because thermal conductivities are not additive. In a composite material the effective thermal conductivity depends on the microstructure and may also be anisotropic (dependent on direction). Many different equations of models have been proposed to predict the k-value of a composite from those of its components. Some are derived from geometric or physical considerations while others are purely empirical. The differences between

Table 2.4 Coefficients of the equation $y = A + B\theta + C\theta^2$ for the dependence of thermal conductivity on temperature (Choi and Okos 1986) and values at −40, −20 and 20 °C. Note: in the original paper there was an order-of-magnitude error in the B-value for fat (M. Okos, personal communication with D. Cleland) which has been corrected here

Component	A	B	C	Value at −40 °C	Value at −20 °C	Value at 20 °C
Water	0.57109	1.7625×10^{-3}	-6.7036×10^{-6}	0.490	0.533	0.604
Ice	2.21960	-6.2489×10^{-3}	1.0154×10^{-4}	2.632	2.385	–
Protein	0.17881	1.1958×10^{-3}	-2.7178×10^{-6}	0.127	0.154	0.202
Fat	0.18071	-2.7604×10^{-4}	-1.7749×10^{-7}	0.191	0.186	0.175
Carbohydrate	0.20141	1.3874×10^{-3}	-4.3312×10^{-6}	0.139	0.172	0.227
Fibre	0.18331	1.2497×10^{-3}	-3.1683×10^{-6}	0.128	0.157	0.207
Ash	0.32962	1.4011×10^{-3}	-2.9069×10^{-6}	0.269	0.300	0.356

prediction methods are particularly acute for frozen foods and porous foods, because the differences between component thermal conductivities can be very great (one order of magnitude for frozen non-porous foods, up to three orders of magnitude for frozen porous foods).

The parallel model (k_{par}) and the series model (k_{ser}), provide the upper and lower bounds of the thermal conductivity of a mixture:

$$k_{par} = \sum_i \phi_i k_i \qquad (2.39)$$

$$k_{ser} = \left(\sum_i \frac{1}{\phi_i k_i} \right)^{-1} \qquad (2.40)$$

where ϕ_i are the volume fraction and k_i the thermal conductivities of pure components. The volume fraction ϕ_i for a given component i can be calculated from

$$\phi_i = \frac{X_i / \rho_i}{\sum_j (X_j / \rho_j)} \qquad (2.41)$$

where ρ_i is the density of component i. Since this work does not aim to make a comprehensive review of thermal properties models but only to provide some basic and useful tools for use in freezing time prediction, only a few thermal conductivity models will be mentioned. These are the Maxwell–Eucken (M-E) model, the Levy model and the effective medium theory (EMT) model. Many other equations are mentioned in the literature, but most of them involve one or more empirical curve-fitting parameters. Some are parameter-free, such as the co-continuous model (Wang et al. 2008), but lack experimental verification at this time.

- M-E model: applies to a dilute dispersion of spheres (component 2) in a continuous phase (component 1):

$$k_{M-E} = k_1 \frac{2k_1 + k_2 - 2(k_1 - k_2)\phi_2}{2k_1 + k_d + (k_1 - k_2)\phi_2} \tag{2.42}$$

- Levy model: Levy (1981) proposed a mathematical modification to the M-E model to eliminate the distinction between disperse and continuous phase, which is not always known.

$$k_{Levy} = k_1 \frac{2k_1 + k_2 - 2(k_1 - k_2)F}{2k_1 + k_2 + (k_1 - k_2)F} \tag{2.43}$$

where

$$F = \frac{1}{2}\left[\frac{2}{G} - 1 + 2\phi_2 - \sqrt{\left(\frac{2}{G} - 1 + 2\phi_2\right)^2 - \frac{8\phi_2}{G}}\right] \tag{2.44}$$

$$G = \frac{(k_1 - k_2)^2}{(k_1 + k_2)^2 + k_1 k_2 / 2} \tag{2.45}$$

It can be seen that the Levy model is the same as the M-E model with the dispersed volume fraction ϕ_2 replaced by the parameter F. It can be shown that the equation is symmetric, that is reversing indices 1 and 2 will produce the same numerical result. Wang et al. (2006) gave Levy's equation a physical basis by showing that it represents the effective conductivity of a dispersion of component 1 in component 2 co-existing with a dispersion of 2 in 1, with equal conductivities and each obeying the M-E model.

- EMT model: for a mixture consisting of n components the effective conductivity k_{EMT} is given by the implicit equation:

$$\sum_{i=1}^{n} \phi_i \frac{k_i - k_{EMT}}{k_i + 2k_{EMT}} \tag{2.46}$$

Of the above models, the M-E and Levy apply to pairs of components while EMT, parallel and series can apply to any number of components. To apply the M-E and Levy models to a multi-component system, we have to combine the components one by one, and the order in which they are introduced may be important.

Figure 2.6 illustrates how the different models predict variations in the thermal conductivity of the composite. It plots k_{eff}/k_1 where k_{eff} is the composite's effective conductivity and k_1 that of the more conducting component, against the volume fraction ϕ_2 of the less conducting component. The first figure is for $k_2/k_1 = 0.1$, a situation which typically arises in a frozen food containing ice ($k \approx 2.2-2.5$ W·m^{-1}K^{-1}),

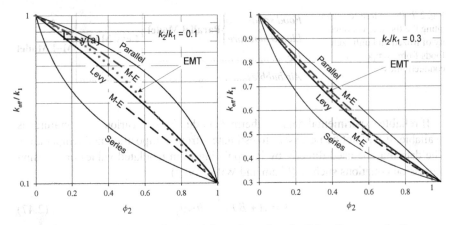

Fig. 2.6 Ratio of effective thermal conductivity to thermal conductivity of more conducting component, as predicted by the parallel, series, Maxwell-Eucken (M-E), EMT and Levy models. The transition between the two parts of M-E is arbitrary

proteins, carbohydrates and fats ($k \approx 0.18 - 0.19$ W·m^{-1}K^{-1}). Because the M-E equation is valid only when one component is dispersed in the other and has a low volume fraction, it is represented by two half-curves, the one on the left having component 2 as the dispersed phase and the one on the right having component 1 as the dispersed phase. The second figure assumes $k_2/k_1 = 0.3$, a situation which typically arises in an unfrozen food ($k \approx 0.6$ W·m^{-1}K^{-1} for water).

If the microstructure of the food is known, the appropriate physical model should be used. For example, for porous food containing air bubbles or emulsions containing droplets, the M-E model is clearly appropriate as long as the volume fraction of the bubbles or droplets is small.

In many cases, however, the microstructure is not known and we have to rely on past experience. The following generic procedure is recommended:

a. For unfrozen foods and the unfrozen phase (liquid water and food solids) of frozen foods, results predicted by the parallel, EMT, M-E and Levy models are fairly close and most workers use the parallel model, which is mathematically the simplest.

b. For frozen foods, the thermal conductivity of the unfrozen phase is calculated as above, then ice and the unfrozen phase are combined using the Levy model (Pham and Willix 1989; Fricke and Becker 2001; Tarnawski et al. 2005; Pham et al. 2006a).

c. For porous foods, the thermal conductivity of the dense (air-free) phase is calculated as above, then the M-E model with air as the dispersed component is used to take into account the effect of air bubbles (Wang et al. 2010).

The procedure is summarised in Fig. 2.7.

In moist porous foods, the thermal conductivity of air bubbles may be augmented by condensation–evaporation (Hamdami et al. 2003). This will be treated further in Sect. 6.2.2 (Eq. 6.23).

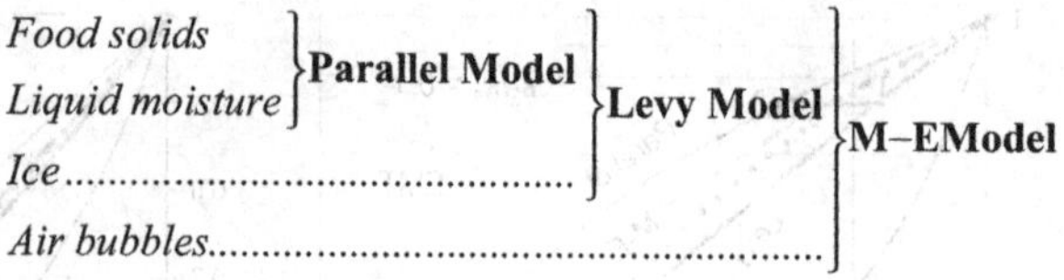

Fig. 2.7 Procedure for calculating the thermal conductivity of frozen and unfrozen
foods from component
conductivities

$$\left. \begin{array}{l} \textit{Food solids} \\ \textit{Liquid moisture} \end{array} \right\} \textbf{Parallel Model} \\ \textit{Ice} \dots\dots\dots\dots\dots\dots\dots\dots\dots\dots\dots\dots\dots\dots\dots\dots\dots\dots\dots \end{array} \right\} \textbf{Levy Model} \\ \textit{Air bubbles} \dots\dots\dots\dots\dots\dots\dots\dots\dots\dots\dots\dots\dots\dots\dots\dots\dots\dots\dots\dots \end{array} \right\} \textbf{M--EModel}$$

If reliable experimental data for thermal conductivity at various temperatures is available, it will not be necessary to estimate k from composition and temperature. The data may be used directly by interpolating from tabulated values or by using regression equations such as (Pham and Willix 1989)

$$k = A + B\theta + \frac{C}{\theta}, \theta < \theta_f \qquad (2.47)$$

$$k = D + E\theta, \theta \geq \theta_f \qquad (2.48)$$

where A–E are regression coefficients. Note the similarity between the above equations with the enthalpy equations, Eq. 2.22 and 2.23, which is due to both thermal conductivity and enthalpy being dependent on the ice fraction.

Empirical equations have been proposed for specific types of foods. For some foods thermal conductivity will depend on the direction of heat flow. For example, the k-value of meat is higher along the fiber than at right angle to it (Heldman and Gorby 1975; Mascheroni et al. 1977; Pham and Willix 1989).

2.6   Thermal Properties of Tylose Gel

A food analogue frequently used in freezing and thawing experiments is Tylose (methyl hydroxyethyl cellulose) gel. Since experiments performed on this material are often used to compare or evaluate freezing time calculation methods, it is important to use accurate property values. Pham and Willix (1990) pointed out that previous estimates of Tylose thermal conductivity by Bonacina and Comini (1971) probably led to erroneous conclusions on the accuracy of finite difference calculations by Cleland et al. (1982). Unfortunately these erroneous data were still repeated in Otero et al. (2006).

The most accurate measurements of thermal conductivity of unsalted MH1000 Tylose gel (77% water mass fraction) were carried out by Pham and Willix (1990) using the guarded hot plate method. Calorimetric properties were determined by Riedel (1960) and Pham et al. (1994) using adiabatic calorimeters. The results are summarised in the following regression equations:

Freezing point:

$$\theta_f = -0.62 \pm 0.05 \, ^\circ C \qquad (2.49)$$

Thermal conductivity:

$$k_u = 0.467 + 0.00154 \ (\theta - \theta_f)$$ (2.50)

$$k_f = 0.467 - 0.00489 \ (\theta - \theta_f) + 0.582 \ (1/\theta - 1/\theta_f)$$ (2.51)

(maximum error 2.5%, mostly due to experimental scatter)
  Specific enthalpy:

$$H(\theta < \theta_f) = A + c_f\theta + B/\theta$$ (2.52)

$$H(\theta \geq \theta_f) = H_0 + c_u\theta$$ (2.53)

where $A = 82.4 \ \text{kJ·kg}^{-1}$, $c_f = 2.31 \ \text{kJ·kg}^{-1}\text{K}^{-1}$, $B = -133.0 \ \text{kJ·K·kg}^{-1}$, $H_0 = 299 \ \text{kJ·kg}^{-1}$, $c_u = 3.78 \ \text{kJ·kg}^{-1}\text{K}^{-1}$, $\theta_f = -0.62 \,°\text{C}$. The maximum error is $10 \ \text{kJ·kg}^{-1}$ except immediately (within 1 K) below the freezing point, where latent heat release causes a steep rise in the enthalpy curve and experimental data is very uncertain, since a small change in temperature will cause a very large change in enthalpy.

Otero et al. (2006) reviewed thermal property data for Tylose gel and proposed equations for extrapolating to higher pressures. Unfortunately some of the data reviewed is out of date and probably inaccurate, especially for thermal conductivity.

2.7 Summary and Recommendations

- The heat transfer coefficient h is frequently the most uncertain parameter in a freezing calculation. However, errors in h are not serious if internal resistance to heat transfer is the controlling factor.
- If reliable experimental data on physical properties is available at various temperatures, they should be used either by direct interpolation from tabulated data or by replacing the data with regression equations. A suitable regression equation for enthalpy below freezing is Eq. 2.22, while above freezing a constant specific heat can usually be assumed (Eq. 2.23). For thermal conductivity, Eq. 2.47 and 2.48 may be used.
- If property data is unavailable but the food's composition is known, the freezing point can be estimated from Eq. 2.11 or similar relationships. The ice fraction at various temperatures can then be estimated from Eq. 2.16. Enthalpy can then be calculated by adding component enthalpies (Eqs. 2.18, 2.19, 2.20) using the specific heat data in Table 2.3. Thermal conductivity can be calculated by the method of Fig. 2.7.
- If only the water content is available, use Eq. 2.25 for enthalpy and Eq. 2.17 for ice fraction. Thermal conductivity can then be estimated by guessing the contents of protein, fat, carbohydrate, fibers and minerals from similar products

and applying the method of Fig. 2.7. Since the thermal conductivities of likely major components except water/ice (which is usually known) and ash (which is usually present in very small amounts) are within a two-to-one range, a reasonable estimate can usually be obtained, say ±10% for non-porous products. Empirical relationships may also be available for specific classes of product (meat, fish, fruit etc.) in the literature.

- Apparent specific heat can in all cases be obtained by numerical or analytical differentiation of the enthalpy–temperature relationship.
- Calorimetric properties at high pressure are readily obtained from those at atmospheric pressure by using basic thermodynamic relationships, provide the values of latent heat of freezing, density and thermal expansion coefficient are available.

⚠ CAUTION

- In Choi and Okos (1986) there was an error in a parameter for fat thermal conductivity, which has been corrected in Table 2.3.
- For water density, Eq. 2.5 is recommended over Choi and Okos's equation.
- For water specific heat, Eq. 2.21 is recommended over Choi and Okos's equation.
- Many papers use obsolete thermal properties for Tylose gel. Those presented in Sect. 2.6 are recommended provided the composition is as described in that section.
- Specific heat may mean sensible heat only or may include latent heat depending on context. Make sure that the correct interpretation is applied.
- The prediction method given above for thermal conductivity may not work well for foods with anisotropic microstructures, such as muscle meat.

Chapter 3
Analytical Solutions

3.1 The Heat Conduction Equation

Food is chilled or frozen by contacting it with a cold medium to remove heat from it. In a solid, the heat transfer is described by Fourier's law of heat conduction, which states that the heat flow per unit area, or heat flux, is proportional to the temperature gradient dT/dx:

$$q_x = -k\frac{dT}{dx}. \tag{3.1}$$

In this equation, q_x is the heat flux (Wm^{-2}K^{-1}), T is the temperature (K), x is the distance in the direction of heat flow (m) and k is the thermal conductivity (Wm^{-1}K^{-1}). If the heat flow direction is not in the x-direction, then the equation is generalized to 3-D:

$$\begin{cases} q_x = -k\dfrac{\partial T}{\partial x} \\[2mm] q_y = -k\dfrac{\partial T}{\partial y} \\[2mm] q_z = -k\dfrac{\partial T}{\partial z} \end{cases} \tag{3.2}$$

or, in vector notation

$$\mathbf{q} = -k\nabla T \tag{3.3}$$

where $\mathbf{q} = (q_x, q_y, q_z)^T$ is the heat flux vector and $\nabla = \left(\dfrac{\partial}{\partial x}, \dfrac{\partial}{\partial y}, \dfrac{\partial}{\partial z}\right)^T$ is the gradient operator. When the heat flux $\mathbf{q}$ is not uniform, heat energy accumulates or depletes locally, causing local temperature to change according to Carslaw and Jaeger 1959 (p. 10)

Q. T. Pham, *Food Freezing and Thawing Calculations*, SpringerBriefs in Food, Health, and Nutrition, DOI 10.1007/978-1-4939-0557-7_3, © The Author 2014

$$\rho c \frac{\partial T}{\partial t} = \frac{\partial}{\partial x}\left(k \frac{\partial T}{\partial x} + k \frac{\partial T}{\partial y} + k \frac{\partial T}{\partial z} \right) + S_q \qquad (3.4)$$

or in vector notation

$$\rho c \frac{\partial T}{\partial t} = \nabla \cdot (k \nabla T) + S_q \qquad (3.5)$$

where ρ is the solid's density (kg·m^{-3}), c its specific heat (J·kg^{-1}K^{-1}) and S_q any in-ternally generated heat (W·m^{-3}) such as that due to microwave radiation. The above equations implicitly assume constant density, since if density varies there will be expansion or contraction and a convection term would have to be added. Methods to deal with variable density will be considered later (Chap. 5). For most freezing operations, there is no internally generated heat, however, in some mathematical modelling methods S_q is used to represent the latent heat release and is therefore non-zero. In other modelling methods, latent heat is not treated separately but is included in the specific heat c (i.e. $c = c_{app}$) in Eq. 3.4 and 3.5.

To solve the heat conduction equation, boundary conditions must be known. The most common one is the convective boundary condition, also known as Newton's cooling law or boundary condition of the third kind, which specifies that the heat flux (W·m^{-2}) at the surface, q_s, is proportional to the difference between the surface temperature T_s and the ambient temperature T_a:

$$q_s = -(k \nabla T)_s \cdot \mathbf{n} = h(T_s - T_a) \qquad (3.6)$$

where $\mathbf{n}$ is the outward normal unit vector and h is the heat transfer coefficient (W·m^{-2}K^{-1}). When h tends to infinity, we obtain the Dirichlet or first-kind bound-ary condition, $T_s = T_a$, and when $h = 0$, we obtain the adiabatic boundary condition. More complex boundary conditions arise if there is also radiative or evaporative heat transfer, as commonly happens.

3.2   Analytical Solutions for Freezing Time

Analytical expressions for the freezing time can be derived for some simple, ide-alised situations. Although usually not directly applicable to practical cases, they are often used to verify numerical models such as finite difference or finite element to ensure that numerical errors are negligible. Cleland (1990) listed more than 100 references presenting analytical solutions, some exact and some approximate, for phase change problems. We shall give a full derivation of Plank's (1913a, b) exact solution because it forms the basis of several important practical expressions for the calculation of freezing times and heat loads.

3.2.1 Solution for Zero Internal Resistance

For a very small object or a well-stirred liquid, or when thermal conductivity is very high, the internal resistance to heat transfer can be neglected and only the external or surface resistance matters. Furthermore, if we neglect sensible heat (heat due to temperature change) and consider only the latent heat, and assume that the latter is released at a constant temperature T_f, the whole object including the surface will be at a constant temperature T_f during freezing. The rate of heat loss across the surface is then $hA(T_f - T_a)$ where A is the surface area (m²). The freezing time is the total latent heat content divided by the heat flow:

$$t = \frac{\rho V L_f}{hA(T_f - T_a)} \qquad (3.7)$$

where ρ is the density (kg·m^{-3}), L_f the latent heat of phase change per unit mass of food (J·kg^{-1}) and V the food's volume. Although this result is trivial, it is listed here for later comparison with more complex situations.

3.2.2 Solution for Zero Sensible Heat: Plank's Equation

Let us consider a solid food in the shape of a slab of constant thickness being cooled by exposing both faces to a cold medium at temperature T_a, which is below its freezing point T_f. The following assumptions are made:

a. The sides of the slab are perfectly insulated, or the slab has infinite area, so that the heat flux is normal to its faces.
b. The phase change takes place at a single sharp temperature T_f, at which latent heat λ (J·kg^{-1}) is released.
c. At the surface, the convective boundary condition (Eq. 3.6) applies.
d. The sensible specific heat of the material is zero both above and below the freezing point.
e. The thermal conductivity k_f of the frozen material is constant.

As the material has no sensible specific heat, it will immediately fall to the freezing point throughout. A freezing front will form at the surface at time $t=0$, then advance towards the centre (Fig. 3.1). The heat conduction equation, Eq. 3.4, can be simplified to

$$0 = \frac{d}{dx}\left(k_f \frac{dT}{dx}\right). \qquad (3.8)$$

Fig. 3.1 Freezing of a slab
with zero sensible heat

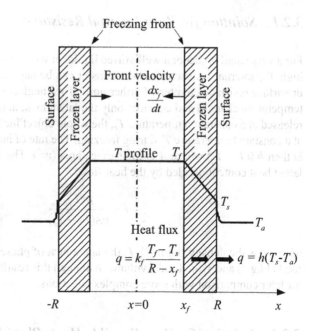

As the time-dependent term in Eq. 3.4 has disappeared, the Plank's solution is termed a quasi-steady state solution. This leads to

$$\frac{dT}{dx} = const \qquad (3.9)$$

i.e. the temperature profile is linear except at the freezing front, where latent heat evolves. In the unfrozen core, there is no heat flow since the sensible specific heat is zero, therefore the temperature must be constant and equal to T_f. For the frozen region, the surface $(x=R)$ is at T_s while the frozen front $(x=x_f)$ is at T_f, hence the temperature gradient just inside the surface will be $(T_s - T_f)/(R-x_f)$. The heat flux q in the frozen layer is therefore $k_f(T_s - T_f)/(R-x_f)$ and it must be equal to the heat flux leaving the surface, $h(T_s - T_a)$:

$$q = k_f \frac{T_f - T_s}{R - x_f} = h(T_s - T_a). \qquad (3.10)$$

Eliminating T_s we get

$$q = \frac{h(T_f - T_a)}{1 + \dfrac{h(R - x_f)}{k_f}}. \qquad (3.11)$$

This heat is generated at the freezing front x_f where latent heat is released. Over a time interval dt the heat released per unit area ($J \cdot m^{-2}$) is qdt and this heat is generated by the freezing of a volume dx_f per unit area, containing latent heat $\rho L_f dx_f$, hence:

$$\rho L_f dx_f = -qdt. \tag{3.12}$$

Substituting for q from Eq. 3.11 we obtain

$$\rho L_f dx_f = -\frac{h(T_f - T_a)}{1 + \dfrac{h(R - x_f)}{k_f}} dt. \tag{3.13}$$

Integrating this equation from $x_f = R$ to 0 gives the total freezing time, t_{Plank}:

$$t_{Plank}(slab) = \frac{\rho L_f}{h(T_f - T_a)}\left(R + \frac{h}{2k_f} R^2 \right). \tag{3.14}$$

We shall call this result the Plank's freezing time. Plank also derived expressions for the freezing times for an infinite cylinder and a sphere under the same assumptions as above. Pham (1986a) showed that Plank's solutions can be written in the following form for the three basic shapes:

$$t_{Plank} = \frac{\rho L_f R}{E_f h(T_f - T_a)}\left(1 + \frac{Bi_f}{2} \right) \tag{3.15}$$

where the shape factor E_f is 1 for slabs, 2 for infinite cylinders and 3 for spheres. Bi_f is the Biot number (based on frozen food thermal conductivity), a very important parameter which will be looked at closely later:

$$Bi_f = \frac{hR}{k_f}. \tag{3.16}$$

3.2.3 The Biot Number

We observe that for the three basic shapes, E_f is equal to the ratio AR/V, where V is the object's volume and A its surface area (both sides for the slab). Substituting for E_f in Eq. 3.15 gives the following expression (Pham 1986a):

$$t_{Plank} = \frac{\rho V L_f}{hA(T_f - T_a)}\left(1 + \frac{Bi_f}{2} \right). \tag{3.17}$$

The first term on the right of the equation is the freezing time for an object with zero internal resistance to heat transfer (Eq. 3.7), while the second term $(1 + Bi_f/2)$

is a correction factor for finite conductivity, or internal resistance. During freezing, the latent heat flux has to traverse a distance varying from 0 when the freezing front coincides with the surface to R when it reaches the centre. The mean distance traversed is therefore $R/2$ and the average internal resistance to heat transfer is $R/2k$. Therefore, $Bi_f/2 = hR/2k_f$ is an estimate of the ratio of internal to external (surface) resistances to heat transfer during the freezing process.

For extreme values of the Biot number, the Plank's freezing time tends to the following limit:

- $Bi_f = \infty$ (negligible surface resistance, e.g. $h = \infty$ or Dirichlet boundary condition):

$$t_{Plank} = \frac{\rho L_f R^2}{2 E_f k_f (T_f - T_a)} \qquad (3.18)$$

i.e. the freezing time is proportional to the square of size.

- $Bi_f = 0$ (negligible internal resistance):

$$t_{Plank} = \frac{\rho L_f R}{E_f h (T_f - T_a)} \qquad (3.19)$$

i.e. the freezing time is proportional to size.

It is essential that practicing engineers be aware of the Biot number of their heating and cooling processes. If it is much larger than 1 (i.e. if internal resistance is controlling heat transfer), then there is little use trying to improve heat transfer coefficient say by reducing the amount of wrapping or increasing air velocity, and the engineer should try to reduce the thickness $2R$ of the product (if possible). On the other hand, when $Bi_f \ll 1$, reducing the thickness will be less effective and efforts should be made to increase the heat transfer coefficient. Note, however, that breaking up the food in smaller packages or pieces will increase the surface area to volume ratio A/V, and hence reduce the freezing time according to Eq. 3.17 no matter what the Biot number is.

3.2.4 Shape Factors for Zero Sensible Heat in Two and Three Dimensions

McNabb et al. (1990a, b) extended Plank's equation to multidimensional regular shapes (infinite rectangular rods, finite cylinders, rectangular bricks). Hossain et al. (1992a) presented these solutions in the form of shape factor E_f, the ratio of freezing time for the given shape to Plank's freezing time for the infinite slab whose thickness is equal to the smallest dimension $2R$:

a. Infinite rectangular rod of sides $2R \times 2\beta_1 R$:

$$E_f = \cfrac{1 + \cfrac{2}{Bi_f}}{\left[\left(1 + \cfrac{2}{Bi_f}\right) - 4\sum_{n=1}^{\infty} \sin z_n \left[z_n^{\,3}\left(1 + \cfrac{\sin^2 z_n}{Bi_f}\right)\left(\cfrac{z_n}{Bi_f}\sinh \beta_1 z_n + \cosh \beta_1 z_n\right)\right]\right]^{-1}}$$

(3.20)

where z_n are the roots of

$$Bi_f = z_n \tan z_n$$

(3.21)

b. Finite cylinder of radius R and height $2\beta_1 R$ (height > diameter):

$$E_f = \cfrac{2 + \cfrac{4}{Bi_f}}{\left[\left(1 + \cfrac{2}{Bi_f}\right) - 8\sum_{n=1}^{\infty}\left[y_n^{\,3} J_1(y_n)\left(1 + \cfrac{y_n^2}{Bi_f^{\,2}}\right)\left(\cosh \beta_1 y_n + \cfrac{y_n}{Bi_f}\sinh \beta_1 y_n\right)\right]\right]^{-1}}$$

(3.22)

where y_n are the roots of

$$y_n J_1(y_n) - Bi_f \cdot J_0(y_n) = 0$$

(3.23)

and J_0, J_1 are the Bessel functions of the first kind of order zero and one respectively.

c. Finite cylinder of radius $\beta_1 R$ and height $2R$ (diameter > height):

$$E_f = \cfrac{1 + \cfrac{2}{Bi_f}}{\left[\left(1 + \cfrac{2}{Bi_f}\right) - 4\sum_{n=1}^{\infty} \sin z_n \left[z_n^{\,2}(z_n + \cos z_n \sin z_n)\left(I_0(\beta_1 z_n) + \cfrac{z_n}{Bi_f} I_1(\beta_1 z_n)\right)\right]\right]^{-1}}$$

(3.24)

where z_n are the roots of Eq. 3.21, and I_0, I_1 are the Bessel functions of the second kind of order zero and one respectively.

d. Rectangular bricks of sides $2R \times 2\beta_1 R \times 2\beta_2 R$ ($\beta_2 > \beta_1 > 1$):

$$E_f = \left(1 + \cfrac{2}{Bi_f}\right)\left[\left(1 + \cfrac{2}{Bi}\right) - 4\sum_{n=1}^{\infty} \sin z_n \left[z_n^{\,3}\left(1 + \cfrac{\sin^2 z_n}{Bi_f}\right)\left(\cfrac{z_n}{Bi_f}\sinh \beta_1 z_n + \cosh \beta_1 z_n\right)\right]\right]^{-1}$$

$$-8\beta_2^{\,2}\sum_{n=1}^{\infty}\sum_{m=1}^{\infty}\cfrac{\sin z_n \sin z_m}{\left(\cosh z_{nm} + \cfrac{z_{nm}}{\beta_2 Bi_f}\sinh z_{nm}\right)z_n z_m z_{nm}^{\,2}\left(1 + \cfrac{\sin^2 z_n}{Bi_f}\right)\left(1 + \cfrac{\sin^2 z_m}{\beta_1 Bi_f}\right)}$$

(3.25)

where z_n are the roots of Eq. 3.21, z_m the roots of

$$\beta_1 Bi_f = z_m \tan z_m \qquad (3.26)$$

and z_{nm} is given by

$$z_{nm}^2 = \beta_2^2 z_n^2 + \frac{\beta_2^2}{\beta_1^2} z_m^2. \qquad (3.27)$$

The roots of Eq. 3.21, 3.23 and 3.26 are tabulated in Carslaw and Jaeger (1959). The shape factor E_f calculated according to Eq. 3.20–3.25 is plotted in Figs. 3.2 and 3.3. Hossain et al. (1992a) gave approximate simplified formulas for E_f, which we will look at in the next chapter. It can be seen from the graphs that E_f tends towards AR/V as Bi tends to 0 (external resistance controlling), but the high Biot number limit is less easy to calculate. In all cases, there is negligible variation in E_f once Bi exceeds about 10.

Approximate expressions for E_f were also derived using the matched asymptotic method (McNabb et al. 1990a, b; Hossain et al. 1992a). Hossain et al. stated that they apply only at low Biot numbers ($Bi_f < 0.16$). Also, they are not accurate at large aspect ratios, since they predict that $E_f \rightarrow 0$ as $\beta_1, \beta_2 \rightarrow \infty$. Therefore, these expressions are of limited usefulness.

3.2.5   Exact Analytical Solutions for Freezing with Sensible Heat

A number of analytical solutions have been obtained without making the zero sensible heat assumption that underlies Plank's equation. In such situations, an important parameter is the Stefan number, Ste, which measures the ratio of the sensible heat for cooling to the latent heat:

$$Ste = \frac{c_f(T_f - T_a)}{L_f} \qquad (3.28)$$

where c_f is the specific heat of frozen food, *not* including latent heat. The larger the Stefan number, the less accurate Plank's equation becomes. In practical food freezing Ste would typically be around 0.1–0.3.

When the Dirichlet boundary condition applies ($T_s = T_a < T_f$) it can be shown by writing the heat conduction equation (Eq. 3.5) in a dimensionless form that the distance of the freezing front from the surface, δ, varies as $\sqrt{\alpha_f t}$ where $\alpha_f = k_f / \rho_f c_f$ is the thermal diffusivity of the frozen phase (Budhia and Kreith 1973). Thus, for the freezing of a semi-infinite body (half space) we obtain Neumann's solution (Carslaw and Jaeger 1959, p. 283):

Fig. 3.2 Analytical freezing shape factors (Hossain et al. 1992a) for regular two-dimensional shapes

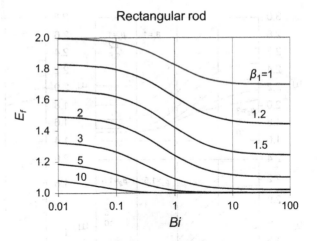

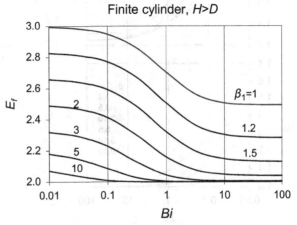

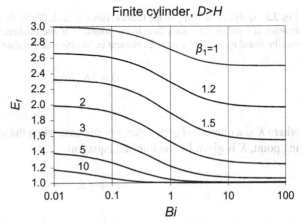

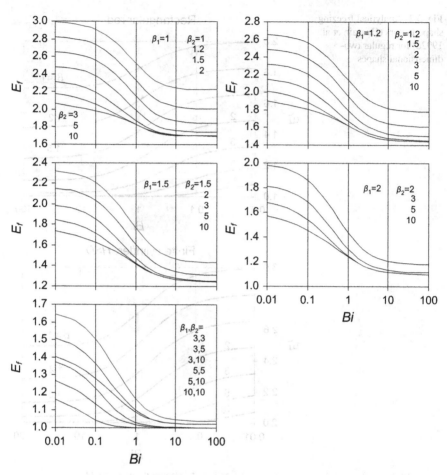

Fig. 3.3 Analytical freezing shape factors (Hossain et al. 1992a) for rectangular bricks. *Curves* are in order of β-parameter values from top to bottom. For intermediate values of β_1, β_2 the value of E_f may be found by interpolation (e.g. bilinear or bicubic interpolation)

$$\delta = 2K\sqrt{\alpha_f t}$$

$$(3.29)$$

where K is a numerical parameter. For the case when the slab is initially at its freezing point, K is given by the implicit equation

$$Ke^{K^2}\,erf\,K = \frac{Ste}{\sqrt{\pi}}$$

$$(3.30)$$

If *Ste* is small the following approximate expression can be used:

$$K \approx \sqrt{\frac{Ste}{2}}.$$

$$(3.31)$$

Neumann's solution can be used to calculate the approximate temperature change, heat load and freezing rate in the initial period of freezing, when the frozen layer is thin compared to the dimension of the product so that surface curvature can be neglected and heat transfer is almost one-dimensional.

The freezing of a wedge (corner region with angle anywhere between 0 and 360°) was solved analytically by Budhia and Kreith (1973). Solutions for hyperboloids (including a cone) melting away from the midplane which is kept at a constant temperature and for ellipsoids and paraboloids growing in a supercooled medium are also available (Ivantsov 1947; Horvay and Cahn 1961; Sekerka and Wang 1999).

3.2.6   Perturbation Solutions for Freezing with Sensible Heat

Approximate solutions have been derived for the freezing of spheres and cylinders with non-zero sensible heat by using perturbation methods (Pedroso and Domoto 1973; Riley et al. 1974). The solutions are in the form of power series of Ste. For example, for a sphere initially at the freezing temperature and subjected to the Dirichlet boundary condition $T_s = T_a$, Riley et al. (1974) derived the equation

$$t_f = \frac{\rho L_f R^2}{k_f (T_f - T_a)} \left[\frac{1}{6} + \frac{Ste}{6} - \frac{Ste^{3/2}}{4\sqrt{2\pi}} + O(Ste^2) \right] \tag{3.32}$$

where $O(Ste^2)$ is an error term whose magnitude is of the order of Ste^2. We observe that when $Ste \to 0$, Eq. 3.32 reduces to Plank's equation (Eq. 3.18), with $E_f = 3$. When $Ste = 0.1$ Riley's freezing time (Eq. 3.32 without the error term) is about 8 % greater than Plank's, and when $Ste = 0.2$ the difference is about 15 %.

3.3   Summary and Recommendations

- The analytical freezing time formulas presented in this chapter are of little practical usefulness due to the unrealistic assumptions made in deriving them.
- However, the freezing technologist or researcher should be thoroughly familiar with Plank's equation, which shows the effect of the major variables (environment temperatures, heat transfer coefficients, food dimensions and thermal properties) and which forms the basis for several approximate and empirical equations that will be of more practical usefulness.
- The Biot number Bi_f measures the relative importance of internal to external resistance to heat transfer. For a process to be efficient it should not be too far from unity. If $Bi_f \gg 1$, internal resistance controls heat transfer and there is little use trying to improve the heat transfer coefficient (say by reducing the amount of wrapping or increasing air velocity).

- Making the product thinner (slicing) will reduce the Biot number. Breaking up the food in smaller packages or pieces will do this and also increase the surface area to volume ratio A/V, and hence reduce the freezing time according to Eq. 3.17.
- The effect of product shape can be represented by a shape factors E_f, the ratio of freezing time of an infinite slab to that of a product with different shape and same minimum dimension. Analytical solutions have been given for the following cases: for infinite cylinders $E_f = 2$, for spheres $E_f = 3$, for finite cylinders and rectangular rods E_f is shown in Fig. 3.2 and for rectangular brick-shaped foods E_f is shown in Figs. 3.3. Although derived with the assumption of zero specific heat, these values are applicable with little error to real-life food-freezing situations.

⚠ **CAUTION**

- Older publications in this field often define the Biot number as $2hR/k$ instead of hR/k. Make sure that the correct interpretation is applied.

Chapter 4
Approximate and Empirical Methods

4.1 Introduction

In this work, approximate methods refer to those that are derived from an approximate model of reality, such as those that divide food freezing process into cooling and phase change stages. Empirical methods are those that involve some empirical parameters obtained by curve-fitting experimental data or numerically generated results. Some methods may include elements of both approaches. Cleland (1990) listed 62 empirical methods, and many more have been proposed since then. The earliest methods are empirical equations for specific types of foods under restricted ranges of conditions. Starting with Cleland (1977) and Cleland and Earle (1977, 1979a, 1979b, 1982, 1984a, 1984b) modern approximate and empirical methods attempt to predict the freezing times for a wide range of foods, taking into accounts variations in thermal properties, geometry and freezing conditions. Most of these start from Plank's analytical solution (Eq. 3.14) and attempt to correct for its unfulfilled assumptions, in particular the non-zero sensible heats above and below the freezing temperature and the gradual phase change. In this work we shall only consider some of most general and best verified methods.

4.2 Freezing Time of 1-D Shapes

4.2.1 Cleland & Earle's Empirical Method

Cleland and Earle (1976, 1977, 1979a, 1979b, 1982, 1984a, 1984b) were the first to take a systematic approach to developing an empirical freezing time prediction method that takes into account the sensible heat effects, backed up by carefully collected data covering most of the range of interest to industry. Cleland and Earle's final correlation (1984a) can be written as follows for the three basic shapes (infinite slabs, infinite cylinders and spheres):

Q. T. Pham, *Food Freezing and Thawing Calculations,*
SpringerBriefs in Food, Health, and Nutrition, DOI 10.1007/978-1-4939-0557-7_4,
© The Author 2014

$$t_f = \frac{\rho \Delta H_{10}}{E_f (T_f - T_a)} \left(\frac{2 P_1 R}{h} + \frac{4 P_2 R^2}{k_f} \right) \left[1 - \frac{1.65 Ste}{k_f} \ln \left(\frac{T_c - T_a}{T_{ref} - T_a} \right) \right] \qquad (4.1)$$

where

$$P_1 = 0.5[1.026 + 0.5808 Pk + Ste(0.2296 Pk + 0.1050)]$$

$$P_2 = 0.125[1.202 + Ste(3.410 Pk + 0.7336)]$$

$$Pk = c_u (T_i - T_f)/\Delta H_{10}$$

$$Ste = c_f (T_f - T_a)/\Delta H_{10}$$

$$T_{ref} = 263.15 \text{ K } (-10°C)$$

and ΔH_{10} is the enthalpy change of the product (J·kg^{-1}) between the freezing point T_f and $-10°C$. ΔH_{10} is used instead of the latent heat of freezing to calculate the Stefan and Plank (Pk) numbers, since the latent heat is very difficult to measure separately from the sensible heat. E_f is the freezing shape factor (1 for slabs, 2 for infinite cylinders, 3 for spheres). Pk and Ste are dimensionless numbers which express the magnitude of precooling and subcooling effects, respectively. These effects are mainly incorporated as empirical correction factors in the parameters P_1 and P_2. A similar approach was used by Hung and Thompson (1983).

4.2.2   Mascheroni & Calvelo's Approximate Method

Mascheroni and Calvelo's (1982) method is the first of the approximate methods as defined in this work. The total freezing time is divided into precooling (from initial product temperature to freezing point), phase change and subcooling (down to the final product centre temperature) periods. The phase change time is calculated from Plank's equation (Eq. 3.15), while the precooling and subcooling times are calcu-lated from analytical expressions (Carslaw and Jaeger 1959) which assume uniform initial temperature and constant thermal properties for each period:

$$t_f = t_{precool} + t_{Plank} + t_{subcool} \qquad (4.2)$$

Because the precooling and subcooling expressions involve infinite series, a com-puter program or a graphical solution is necessary for those periods. For this reason the method is not widely used, but its methodology of adding cooling and phase change times forms the basis for simpler methods such as those of Pham (1984,

1986), Ilicali and Saglam (1987) and Lacroix and Castaigne (1988). Two of these, Pham's methods (1984, 1986), will be described in detail next.

4.2.3 Pham's Method 1 (1984)

Pham (1984) proposed the expression

$$t_f = \frac{\rho R}{E_f h}\left[\frac{\Delta H_1}{\Delta T_1}\left(1 + \frac{Bi_1}{3}\right) + \frac{L_f}{\Delta T_2}\left(1 + \frac{Bi_2}{2}\right) + \frac{\Delta H_3}{\Delta T_3}\left(1 + \frac{Bi_3}{3}\right)\right] \qquad (4.3)$$

where

$$\Delta H_1 = c_u(T_i - T_{fav})$$

$$\Delta H_3 = c_f(T_{fav} - T_e)$$

$$\Delta T_1 = \frac{(T_i - T_a) - (T_{fav} - T_a)}{\ln\dfrac{T_i - T_a}{T_{fav} - T_a}}$$

$$\Delta T_2 = T_{fav} - T_a$$

$$\Delta T_3 = \frac{(T_{fav} - T_a) - (T_e - T_a)}{\ln\dfrac{T_{fav} - T_a}{T_e - T_a}}$$

$$Bi_1 = \frac{1}{2}\left(\frac{hR}{k_u} + \frac{hR}{k_f}\right)$$

$$Bi_2 = Bi_3 = \frac{hR}{k_f}$$

T_{fav} is the mean freezing temperature or centre-of-mass of the area under the latent heat peak (see Fig. 2.2), and for water-rich foods, under normal commercial freezing conditions, it is given approximately by

$$T_{fav} = T_f - 1.5$$

T_e is the average product temperature at the end of the process and can be estimated by assuming a linear temperature profile, giving:

$$T_e = T_c - \frac{T_c - T_a}{2 + \dfrac{4}{Bi_3}}$$

The above expression for T_e strictly holds only for slab but for simplicity it is also used for other geometries. All temperatures can be in K or °C. It can be seen that Eq. 4.3 is a form of Eq. 4.2 with the precooling and subcooling times given by approximate analytical expressions. As with Mascheroni and Calvelo's method, Pham's (1984) method does not contain any parameter obtained by curve-fitting freezing time data and can be considered therefore an approximate method, not an empirical one. The use of log-mean temperature differences, mean freezing temperature, mean product temperature and the terms $Bi_1/3$, $Bi_3/3$ for cooling are all derived from analytical considerations.

4.2.4   Pham's Method 2 (1986)

Pham's (1986a) method can be written as follows:

$$t_f = \frac{\rho R}{E_f h} \left(\frac{\Delta H_1}{\Delta T_1} + \frac{\Delta H_2}{\Delta T_2} \right) \left(1 + \frac{Bi_f}{2} \right) \qquad (4.4)$$

where ΔH_1 and ΔT_1 are the specific enthalpy change and temperature difference respectively for the precooling period, and ΔH_2 and ΔT_2 those for the combined freezing–subcooling period, which may be calculated from:

$$\Delta H_1 = \rho c_u (T_i - T_{fm})$$

$$\Delta T_1 = \frac{T_i + T_{fm}}{2} - T_a$$

$$\Delta H_2 = \rho[L_f + c_f (T_{fm} - T_c)]$$

$$\Delta H_2 = \rho[L_f + c_f (T_{fm} - T_c)]$$
$$\Delta T_2 = T_{fm} - T_a$$

T_{fm} is a nominal "mean freezing temperature" (defined differently from T_{fav} in Pham's 1984 method) and the following equation was proposed for most water-rich biological materials:

$$T_{fm} - T_0 = \theta_{fm} = 1.8 + 0.263\theta_c + 0.105\theta_a \qquad (4.5)$$

It can be readily seen that Eq. 4.4 is an extension to Plank's equation (Eq. 3.15), with the term $\Delta H_1/\Delta T_1$ representing the precooling time, $\Delta H_2/\Delta T_2$ representing the

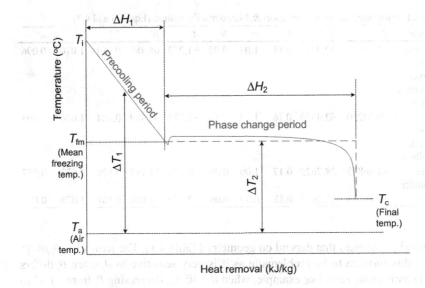

Fig. 4.1 Illustration of Pham's (1986) freezing time equation

phase change–subcooling time, and the term $1 + Bi_f/2$ representing the effect of internal resistance to heat transfer. T_{fm} represents some kind of time-averaged product temperature during the combined phase change and subcooling period. In the expression for T_{fm}, the term involving θ_c expresses the effect of the subcooling period on the mean product temperature, while the term involving θ_a can be interpreted as a correction for the temperature profile in the frozen product during the phase change and subcooling periods. The three constants in T_{fm} are the only empirical parameters in Pham's (1986) method. Figure 4.1 illustrates the physical reasoning behind the method.

Equation 4.4 has been extended to the freezing of foods with variations in environmental conditions (Pham 1986a) and to the asymmetric freezing of slabs (Pham 1987a).

4.2.5 Salvadori & Mascheroni's Method

Salvadori and Mascheroni (1991) proposed the following empirical correlation, obtained by regression of numerically generated temperature profiles:

$$t_f = \left(A\theta_c + B\right)\left(\frac{1}{Bi_u} + C\right)\left(1 - \frac{\theta_i}{\theta_f}\right)^n \left(\frac{\theta_a}{\theta_f} - 1\right)^{-m} \frac{R^2}{\alpha_u} \qquad (4.6)$$

where $\alpha_u = k_u/\rho_u c_u$ is the thermal diffusivity of the unfrozen food, $Bi_u = hR/k_u$ is the Biot number based on unfrozen food thermal conductivity, and A, B, C, m and n are

Table 4.1 Parameter values for Salvadori & Mascheroni's method (Eqs. 4.6 and 4.7)

Geometry	A	B	C	m	N	A'	B'	C'	m'	n'
Slab, heat flow perpendicular to fibres	−1.08125	62.9375	0.18	1.04	0.09	−1.272	65.489	0.184	1.070	0.096
Slab, heat flow parallel to fibres	−0.94250	62.4350	0.16	1.03	0.10	−1.272	65.489	0.184	1.070	0.096
Infinite cylinder	−0.46875	28.7625	0.17	1.00	0.09	−0.750	32.198	0.179	1.032	0.037
Sphere	−0.16875	15.3625	0.18	0.90	0.06	−0.439	24.804	0.167	1.078	0.073

empirical parameters that depend on geometry (Table 4.1). The term $(1 - \theta_i/\theta_f)^n$ causes this formula to be problematic as it is very sensitive to θ_f when θ_i differs significantly from zero. For example, when $\theta_i = 30\,°C$, decreasing θ_f from $-1\,°C$ to $-2\,°C$ cause the freezing time to almost halve, but when $\theta_i = 1\,°C$, the same change in freezing point causes no change in freezing time.

Perhaps because of this inconsistency, these authors later (Salvadori 1994) simplified the equation by assuming $\theta_f = -1\,°C$, leading to

$$t_f = \left(A'\theta_c + B'\right)\left(\frac{1}{Bi_u} + C'\right)\left(1 + \theta_i\right)^{n'}\left(-\theta_a - 1\right)^{-m'}\frac{R^2}{\alpha_u} \qquad (4.7)$$

Unlike other methods, Eqs. 4.6 and 4.7 do not require frozen food properties to be known or estimated (these are implicitly assumed to be uniquely related to the unfrozen food properties) and is therefore easier to apply than other prediction methods. However, since the latent heat and other frozen food properties do not appear explicitly in the relationship, Eq. 4.7 could be unreliable if thermal properties differ significantly from those of the foods tested, for example for foods with low or high solute contents, whose freezing points significantly differ from $-1\,°C$. The initial temperature must be greater than $0\,°C$. At initial temperatures around $0\,°C$ the freezing time predicted by Salvadori & Mascheroni's method is very sensitive to initial temperature. For example, changing θ_i from $0\,°C$ to $1\,°C$ causes the predicted freezing time to change by up to 7%, which is clearly unrealistic. Under typical freezing conditions numerical calculations predict a 1.0% change (Fig. 4.2).

4.2.6 Improved Shape Factors for Basic Geometries

Cleland & Earle's and Pham's prediction methods assume that the freezing times for slabs, infinite cylinders and spheres are proportional to the ratio V/AR, that is $E_f = AR/V = V = 1, 2$ and 3, respectively. This is analytically exact for the conditions

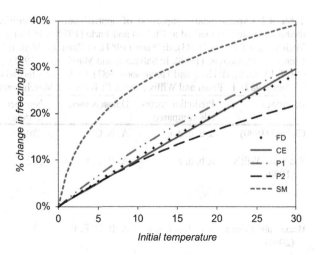

Fig. 4.2 Effect of initial temperatures on freezing time predicted by various methods: finite differences (FD), Cleland and Earle (CE), Pham 1984 (P1), Pham 1986a (P2), Salvadori & Mascheroni (SM). Parameters: $\theta_f = 0.7\,°C$, $Bi_f = 1$, $\theta_A = -40\,°C$

underlying Plank's equation (Eq. 3.15), that is when sensible heat is ignored, and is also valid when internal resistance is negligible ($Bi_f = 0$), but may lead to errors when sensible heat effects are appreciable and the Biot number is large. Ilicali et al. (1999) therefore proposed the following empirical relationships:

- For infinite cylinders:

$$E_f = 1.9621 - 0.0104\theta_c + 0.0015\theta_i + 0.0045\theta_a + \frac{0.0112}{2Bi_f} \tag{4.8}$$

- For spheres:

$$E_f = 2.8842 - 0.0271\theta_c + 0.00457\theta_i + 0.0113\theta_a + \frac{0.0341}{2Bi_f} \tag{4.9}$$

These equations were derived for $0.1 \leq Bi_f \leq 100$. For $Bi_f \geq 100$ the equations remain usable since they tend to asymptotic values. However, for $Bi_f < 0.1$ (negligible internal resistance) the values $E_f = 2$ for infinite cylinders and $= 3$ for spheres should be used, as required by Eq. 3.7.

4.2.7 Comparison of Approximate and Empirical Freezing Time Prediction Methods in 1-D

Several works have compared the accuracy of approximate and empirical relationships, using experimental data and results generated by rigorous numerical methods. Table 4.2 lists some works that compare different methods using composite datasets from several sources and Table 4.3 shows the results for the best perform-

Table 4.2 Experimental comparison of approximate and empirical freezing time prediction methods. Prediction method a: Cleland and Earle (1984b), b: Hung and Thompson (1983), c: de Michelis and Calvelo (1983), d: Pham (1984), e: Pham (1986a), f: Ilicali and Saglam (1987), g: Lacroix and Castaigne (1988), h: Salvadori and Mascheroni (1991). Dataset A: Cleland and Earle (1977, 1979a, b), B: Hung and Thompson (1983), C: de Michelis and Calvelo (1983), D: Ilicali and Saglam (1987), E: Pham and Willix (1990), F: Tocci and Mascheroni (1994)

Reference	Prediction methods compared	Datasets used	Number of data	Comments
Cleland (1990)	a, b, c, d, e	A, B, C	280	Include data for multi-dimensional shapes
Pham and Willix (1990)	a, b, d, e	A, B, C, E	312	Include data for multi-dimensional shapes. Use updated thermal properties for Tylose gel
Becker and Fricke (2004)	a, b, d, e, f, g, h	A, B, D, E, F	274	Include data for non-ideal shapes (apples representing spheres)

Table 4.3 Error statistics for some freezing time prediction methods (mean % error ± standard deviation). (a) Comparison by Cleland (1990), (b) comparison by Pham and Willix (1990), (c) comparison by Becker and Fricke (2004)

Prediction method	Slab	Infinite cylinder	Sphere	All geometries
Cleland & Earle	$+5.6 \pm 5.0$ (c)	$+2.4 \pm 1.7$ (c)	$+9.9 \pm 9.9$ (c)	$+0.2 \pm 6.3$ (a) $+5.1 \pm 8.1$ (b)
Pham 1 (1984)	$+5.9 \pm 4.7$ (c)	$+4.4 \pm 2.9$ (c)	$+12.3 \pm 12.5$ (c)	$+0.4 \pm 7.1$ (a) $+2.7 \pm 7.0$ (b)
Pham 2 (1986)	$+6.6 \pm 5.0$ (c)	$+3.9 \pm 2.8$ (c)	$+10.4 \pm 12.4$ (c)	$+1.1 \pm 7.3$ (a) $+2.1 \pm 6.7$ (b)
Salvadori & Mascheroni	$+7.3 \pm 5.6$ (c)	$+7.3 \pm 5.6$ (c)	$+7.5 \pm 6.9$ (c)	–

ing methods. Statistical criteria for comparison usually include mean error, error range, standard deviation and range enclosing 90 % of data, all in percentage terms. Other criteria may be also evaluated: correlation with exact numerical predictions, kurtosis and skewness of errors. Many other works have evaluated the equations but are not listed here because they use more restricted datasets and/or did not carry out a comparison of methods; however their conclusions are largely similar to those shown.

Some caution must be applied in interpreting these comparisons. The accuracy of prediction methods depends on the values of thermal properties, which vary strongly with temperature below the freezing range (Chap. 2). Since approximate and empirical methods assume constant specific heat and thermal conductivity for the frozen range, some uncertainty in these values is involved. When non-ideal shapes (such as apples used to represent spheres) are included in the evaluation this adds to the errors. Finally, even with the best care in the world it is not easy to ensure that ideal one-dimensional heat flow can be reproduced to an acceptable level in the lab.

Although minor variations exist between the comparisons, some general conclusions can be drawn. On the whole, Cleland & Earle's method, Pham's two methods and Salvadori & Mascheroni's method all produce predictions of acceptable accuracy for most routine industrial applications. All methods except Salvadori & Mascheroni's tend to overpredict the freezing time of spheres.

It must also be kept in mind that the empirical methods were derived based on data for water-rich fresh foods and food analogues (meat, fish, Tylose gel). Their accuracy for processed foods and foods with low water content has not been systematically verified. This limitation will be considered in detail in Sect. 4.2.8.

Regarding the effect of shape, the assumption that the freezing times for slabs, infinite cylinders and spheres are in the ratio $3:2:1$ (i.e. $E_f = 1$, 2 and 3, respectively) may cause an error in calculating the freezing time of cylinders and spheres. Numerical calculations show that E_f may differ from the above values by up to 5% for infinite cylinders and 10% for spheres under usual freezing conditions. Salvadori & Mascheroni's approach allows the effect of shape to be taken into account more precisely since it uses a different set of parameters for each basic shape, hence this method performs best for spheres (Becker and Fricke 2004). Another approach is to employ more accurate empirical expressions for the shape factors of infinite cylinders and spheres, such as Eqs. 4.8 and 4.9.

4.2.8 Freezing Time Prediction for Extended Parameter Range

There are serious gaps in the experimental datasets used to derive or test freezing time prediction methods. Firstly, they were mostly obtained with high moisture content foods, about 70% water or more (meat, fish, fruit). Secondly, the majority of experiments were not even done on food, but on methylcellulose gel (Tylose), a food analogue with supposedly similar properties to meat. Thirdly, due to the cost in time and money of performing experiments, not all combinations of parameters could be covered so there are lots of gaps in the data. There are few tests at very high Biot number, and practically no tests for low moisture foods or those with low freezing point (below about $-1.5\,°C$) due to added salts, such as salted butter or ham. Fourthly, there has been no systematic experiment in the cryogenic temperature range.

To remedy some of these limitations, a factorial experiment using the rigorous finite difference method (Chap. 5) was carried out by Pham (2014) to calculate the freezing times of a slab over the parameter ranges listed in Table 4.4. Product properties were calculated from composition according to the methods of Chap. 2. For Cleland & Earle's and Pham's methods values of frozen thermal conductivity and specific heat are required, but it is unclear at what temperature these should be evaluated, because they vary continuously and significantly with temperature. The methods were therefore tested with frozen properties evaluated at different temperatures and $-40\,°C$ was found to be a suitable temperature for calculating k_f and c_f while $20\,°C$ was used for unfrozen properties. Because the tests included

Table 4.4 Parameter range used in Pham's (2014) numerical experiment to evaluate the agreement between numerical freezing time predictions and those of approximate and empirical formulas

Variable	Range
Water mass fraction, x_w	0.5 to 0.9
Mineral mass fraction, x_m	$(0.1$ to $0.2) \times (1 - x_w)$
Protein mass fraction, x_p	$(0.6$ to $1) \times (1 - x_w - x_m)$
Carbohydrate mass fraction, x_c	0
Fat mass fraction, x_f	$1 - x_w - x_m - x_p$
Initial freezing point (calculated from composition), T_f	-0.5 to $-9.3\,^\circ$C
Initial temperature, θ_i	0 to $40\,^\circ$C
Environment temperature, θ_a	-55 to $-25\,^\circ$C
Heat transfer coefficient, h	1 to 64 W·m^{-2}K^{-1}
Slab half-thickness, R	0.1 m
Biot number, Bi_f	0.045 to 31
Final centre temperature, θ_c	$-18\,^\circ$C

Table 4.5 Comparison of freezing times predicted by finite difference method (FD) with those predicted by approximate and empirical methods. Error = (formula prediction − FD prediction)/ FD prediction × 100%

	Cleland & Earle	Pham 1 (1984)	Pham 2 (1986)	Salvadori & Mascheroni
Mean	23%	−6%	−5%	−12%
S.d.	34%	6%	13%	12%
Minimum	−7%	−23%	−43%	−53%
Maximum	219%	6%	19%	21%
Correlation with FD	0.9823	0.9994	0.9869	0.9958

foods with freezing points down to $-9.3\,^\circ$C, ΔH_{10} in Cleland & Earle's method could miss most of the latent heat and was therefore replaced by the enthalpy change from T_f to $T_f - 10$. The latent heat L_f used in Pham's Method 2 was calculated from

$$L_f = H(\theta_f) - H(-40^\circ C) - c_f(\theta_f + 40) \qquad (4.10)$$

Results of the comparison are shown in Table 4.5. Errors for Cleland & Earle's methods are much larger than previously reported. The large discrepancies occur at combinations of low water content ($x_w = 0.5$) and Biot numbers larger than 1 (when the effect of internal resistance becomes important), which were not present in the experimental datasets used to derive this empirical method. Pham's Method 1 shows the best error range and correlation with numerical results, which is consistent with the fact that this method was derived (albeit approximately) from basic principles while the others involve many empirical regression parameters.

The error range for even the best method is still unacceptable, so Pham (2014) proposed simple correction factors for Pham's methods to yield improved predictions:

Table 4.6 Error statistics for Pham's corrected methods. Error = (formula prediction − FD prediction)/FD prediction × 100%

	Pham 1 corrected	Pham 2 corrected (Eq. 4.11)	Mean Pham 1&2 corrected (Eq. 4.12)
Mean ± s.d.	0.0%±3.0%	0.1%±3.4%	0.0%±2.5%
Range	−10%, +10%	−11%, +10%	−9%, +9%
5- & 95-percentile	−5%, +5%	−5%, +5%	−4%, +4%
Correlation with FD	0.9996	0.9993	0.9996

$$t_f = \left[1 + 0.41 R_T^{0.5}\left(1 - e^{-Bi_f}\right)\right] t_{f,P1} \tag{4.11}$$

$$t_f = \frac{1}{1.13 + 0.115 e^{-Bi_f} - 0.805 R_T^{0.5}} t_{f,P2} \tag{4.12}$$

where $t_{f,P1}$ is the freezing time predicted by Pham's Method 1 (Eq. 4.3) and $t_{f,P2}$ that predicted by Pham's Method 2 (Eq. 4.4). R_T is a temperature ratio that measures the effect of the freezing point depression and consequent smearing of the latent heat peak over the temperature range (see Figs. 2.2 and 2.4):

$$R_T \equiv \frac{\theta_f}{\theta_a} = \frac{T_0 - T_f}{T_0 - T_a} \tag{4.13}$$

Pham's (2014) tests showed that Eqs. 4.11 and 4.12 yield predictions that agree with finite differences to within about ±10%. Taking the average of these two formulas yields even better agreement with finite differences (Table 4.6). Further tests show that the corrected formulas remain accurate at θ_a down to −105 °C and θ_c down to −30 °C. It should be noted, however, that at cryogenic temperatures secondary effects such as supercooling, vitrification or cracking may become important and the numerical model of this paper may not be reliable. In the range $\theta_f \geq -1.5$ °C and $\theta_a \leq -25$ °C, which apply to practically all previous food freezing experiments and the industrial freezing of all fresh foods, the correction factor for Pham's Method 1 is within 5% of unity, so this method can be used "as is" without the correction factor.

In conclusion, Pham's corrected method 1 (Eqs. 4.3 and 4.11), Pham's corrected method 2 (Eqs. 4.4 and 4.12) or the average of their predictions should be used whenever possible. For fresh, water rich foods, the uncorrected versions (Eq. 4.3 or 4.4) may be used. If thermal properties below freezing and composition data are lacking then Salvadori & Mascheroni's method (Eq. 4.6) can be used for fresh, water rich foods, however its limitations should be kept in mind (sensitivity to initial product temperature, poor accuracy for foods with low moisture or low freezing point).

4.3   Freezing Time of Multidimensional Shapes

All analytical, approximate and empirical freezing time calculation methods start
with simple one-dimensional shapes: slabs, infinite cylinders and spheres. However, industrial products invariably have multidimensional shapes, which may be
regular such as cartons (rectangular bricks) or cylindrical containers, or highly irregular such as meat carcasses. This section considers how the existing calculation
methods can be extended to those geometries.

4.3.1   Equivalent Shape Approach

The freezing of a multidimensional shape can sometimes be approximated by that
of an equivalent basic shape; for example, Ilicali and Holacar (1990) approximated
an ellipsoid by an equivalent sphere with radius intermediate between that of a
sphere with the same surface-to-area ratio and that of a sphere with the same volume. This approach is likely to be useful only when the deviation from the standard
shape not too large.

4.3.2   EHTD Shape Factor Approach

Based on Plank's equations, Cleland and Earle (1982) extended the shape factor
E_f in Eq. 3.15, which they termed the "equivalent heat transfer dimensionality"
(EHTD), to cover arbitrary shapes using the slab as a basis:

$$t_f = t_{f,slab}/E_f \qquad (4.14)$$

For cases where sensible heat can be ignored and latent heat is released at a single
temperature, analytical expressions for E_f were derived by McNabb et al. (1990a,
b) for regular multidimensional shapes (Eqs. 3.20–3.25). Empirical equations for
shape factors for common cases have also been proposed and are listed below. It
should be kept in mind that these tend to be more accurate at moderate aspect ratios
and low to moderate Biot numbers.

4.3.2.1   Infinite Rectangular Rods, Bricks, Finite Cylinders

Hossain et al. (1992a, b, c) verified that the analytical shape factors derived for
Plank's equation (Eqs. 3.20–3.25) succeed in predicting the freezing time for regular multidimensional shapes. However, because they involve infinite series, their
calculation is not straightforward. An alternative is to use the empirical calculation
method by Cleland et al. (1987a, b), which is slightly less accurate:

Table 4.7 Geometric constants for calculating the freezing shape factors E_f of regular multidimensional shapes. (Cleland et al. 1987a, b)

Shape	G_1	G_2	G_3
Finite cylinder, height < diameter	1	2	0
Finite cylinder, diameter < height	2	0	1
Rectangular rod	1	1	0
Rectangular brick	1	1	1

$$E_f = G_1 + G_2 E_1 + G_3 E_2 \tag{4.15}$$

where G_1, G_1 and G_3 are constants that depend on the type of shape, as listed in Table 4.7, while E_1 and E_2 are empirical functions of Bi_f and the aspect ratios β_1, β_2 as defined in the previous chapter:

$$E_1 = X(2.32\beta_1^{-1.77})\frac{1}{\beta_1} + \frac{1-X(2.32\beta_1^{-1.77})}{\beta_1^{1.47}}\frac{0.73}{\beta_1^{2.50}}$$

$$E_2 = X(2.32\beta_2^{-1.77})\frac{1}{\beta_2} + \frac{1-X(2.32\beta_2^{-1.77})}{\beta_2^{1.47}}\frac{0.50}{\beta_2^{3.69}}$$

where $X(x)$ is a function defined by

$$X(x) = \frac{x}{Bi_f^{1.34} + x}$$

4.3.2.2 Elliptical Cylinders (2-D)

McNabb et al. (1990b) derived the following approximate expression, which is accurate for $\beta \leq 10$ (Hossain et al. 1992b) where β is the ratio of larger to smaller dimension:

$$E_f = 1 + \frac{1+2/Bi_f}{\beta^2 + 2\beta/Bi_f} \tag{4.16}$$

Pham's (1991) equation for 3-D ellipsoids (Eq. 4.21) can also be used, by setting $\beta_2 = \infty$ and $\beta_1 = \beta$:

$$E_f = 1 + \left(\frac{AR}{V} - 1\right)^p \beta^{-q-p} \tag{4.17}$$

where

$$p = \left(1 + \frac{Bi_f}{2}\right)^{-1}$$

$$q = \frac{1 + Bi_f/4}{1 + Bi_f/8}$$

Per unit length $V = \pi \beta R^2$, while A can be calculated by an approximate formula such as Ramanujan's approximation

$$A \approx 2\pi R\left[3(1+\beta) - \sqrt{(3+\beta)(1+3\beta)}\right] \tag{4.18}$$

resulting in

$$\frac{AR}{V} \approx \frac{2}{\beta}\left[3(1+\beta) - \sqrt{(3+\beta)(1+3\beta)}\right] \tag{4.19}$$

Ilicali et al. (1999) suggested the following correction:

$$E_f = E_{Pham} - 0.119 - 0.0062\theta_c + \frac{0.01885}{\beta Bi_f} \tag{4.20}$$

where E_{Pham} is the value calculated by Eq. 4.17.

4.3.2.3 Three-Dimensional Ellipsoids

Pham (1991) proposed the following relationship for an ellipsoid with semi-axes $R \times \beta_1 R \times \beta_2 R$:

$$E_f = 1 + \left(\frac{\frac{AR}{V} - 1}{\beta_1^{-1} + \beta_2^{-1}}\right)^p \left(\beta_1^{-q} + \beta_2^{-q}\right) \tag{4.21}$$

where p and q are as given in Eq. 4.17. The volume V of an ellipsoid is given by $V = \frac{4}{3}\pi \beta_1 \beta_2 R^3$, while the calculation of the surface area of an ellipsoid is complicated and involves elliptic integrals. However, simple exact solutions are available for $\beta_1 = 1$ or $\beta_1 = \beta_2$ (Hossain et al. 1992c) and quite accurate approximate formulas exist for the general case, such as the following which is accurate to about 1% (Michon 2013):

$$A \approx 4\pi \left(\frac{a^n b^n + a^n c^n + b^n c^n}{3}\right)^{1/n} \tag{4.22}$$

where $a = R$, $b = \beta_1 R$, $c = \beta_2 R$ and $n = 1.6075$. This approximation leads to

$$\frac{AR}{V} \approx \frac{3}{\beta_1 \beta_2} \left(\frac{\beta_1^n + \beta_2^n + \beta_1^n \beta_2^n}{3} \right)^{1/n} \qquad (4.23)$$

For $\beta_1, \beta_2 \leq 2$, Pham showed that the following equation is accurate to within $\pm 5\%$:

$$E_f = 1 + \beta_1^{-q} + \beta_2^{-q} \qquad (4.24)$$

An alternative approach is to use the three-dimensional version of McNabb et al.'s equation, which is accurate to within about -7% to $+5\%$ (Hossain et al. 1992c) for $\beta_1, \beta_2 \leq 4$:

$$E_f = 1 + \frac{1 + 2/Bi_f}{\beta_1^2 + 2\beta_1/Bi_f} + \frac{1 + 2/Bi_f}{\beta_2^2 + 2\beta_2/Bi_f} \qquad (4.25)$$

4.3.2.4   Irregular Shapes

Irregular shapes can often be approximated by ellipses or ellipsoids, and the shape factors for these can be applied using suitable values for β_1 and β_2. Cleland et al. (1987a, b) and Hossain et al. (1992a, b) recommended using Eq. 4.16 or 4.25, with

$$\beta_1 = \frac{S_1}{\pi R^2} \qquad (4.26)$$

$$\beta_2 = \frac{S_2}{\pi R^2} \qquad (4.27)$$

where S_1 and S_2 are the cross-section areas defined as follows: for 2-D irregular shapes, S_1 is the area and R is the smallest half-dimension; for 3-D irregular shapes, S_1 is the area of the smallest cross-section and S_2 that of the cross-section through the smallest diameter ($2R$) that is orthogonal to the first (Fig. 4.3).

Cleland (1991) recommended using Eq. 4.25 but replacing the object by a model ellipsoid with same characteristic dimension R, same smallest cross-section area S_1 containing R and same volume V. This leads to the following expression:

$$\beta_2 = \frac{V}{\frac{4}{3} \pi \beta_1 R^3} \qquad (4.28)$$

while β_1 is still calculated from Eq. 4.26. Cleland stated that the equation can also be used to estimate the freezing and thawing times of regular shapes (bricks and finite cylinders).

In applying any of these methods, some subjective judgment has to be used for highly irregular or non-symmetrical shapes, and thin protrusions should be ignored since they have negligible influence on the bulk of the product. The geometry may

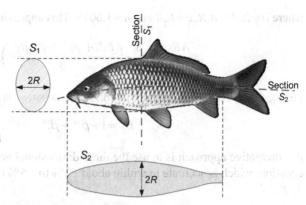

Fig. 4.3 Characteristic dimension R of an irregular shape and orthogonal cross-sections containing R. Notice how thinner sections or protrusions (fins, tail) are ignored as they have negligible effect on the freezing time. Also, the fish would have to be straightened longitudinally to give the cross-section S_2

have to be straightened or otherwise adjusted, since one would expect that a bent rod have a similar freezing time to a straight rod with same cross-section and length.

4.3.3 Mean Conducting Path (MCP) Approach

In Chap. 3 we saw that Plank's solution (Eq. 3.15) can be written in the form of Newton's cooling law (Eq. 3.7) with a correction for the effect of internal resistance, expressed by the term $Bi_f = hR/k_f$ for the slab, infinite cylinder or sphere. R is the distance from the surface to the centre. For other shapes, points on the surface are at different distances from the centre. Pham (1985a) therefore defined a distance termed the mean conducting path or MCP; R_m, which is the mean or effective distance from surface to centre (Fig. 4.4). The effective Biot number then becomes

$$Bi_{fm} = \frac{hR_m}{k_f} = K_R Bi_f \tag{4.29}$$

where K_R is the ratio of MCP to the minimum distance from surface to thermal centre:

$$K_R = \frac{R_m}{R} \tag{4.30}$$

Plank's equation (Eq. 3.15) then becomes, for all shapes:

$$t_{Plank} = \frac{\rho V L_f}{hA(T_f - T_a)}\left(1 + K_R \frac{Bi_f}{2}\right) \tag{4.31}$$

Similarly, Pham's (1986) equation, Eq. 4.4, can be written as

$$t_f = \frac{\rho V}{hA}\left(\frac{\Delta H_1}{\Delta T_1} + \frac{\Delta H_2}{\Delta T_2}\right)\left(1 + K_R \frac{Bi_f}{2}\right) \tag{4.32}$$

Fig. 4.4 Illustration of Mean Conducting Path concept in a rectangular cross-section

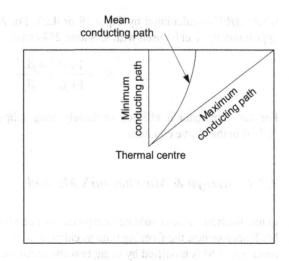

For the three basic shapes $K_R = 1$, while for other shapes K_R must be found by analysis or regression.

4.3.3.1 MCP for Rectangular Rods and Bricks

Pham (1985a) proposed

$$K_R = \frac{R_m}{R} = 1 + \left\{ \left(1.5\sqrt{\beta_1} - 1\right)^{-4} + \left[\left(\frac{1}{\beta_1} + \frac{1}{\beta_2}\right)\left(1 + \frac{2}{Bi_f}\right) \right]^{-4} \right\}^{-1/4} \quad (4.33)$$

(For infinite rods, $1/\beta_2 = 0$.) Pham found that the largest dimension has little influence on the freezing time and therefore also proposed a simpler equation:

$$K_R = \frac{R_m}{R} = 1.43\sqrt{\beta_1} \quad (4.34)$$

This expression may predict a freezing time for the brick that is longer than that of a slab of equal thickness, i.e. $E_f < 1$, therefore the slab freezing time should be used as an upper limit.

4.3.3.2 MCP for Ellipses and Ellipsoids

For 3-D ellipsoids with semiaxes $R \times \beta_1 R \times \beta_2 R$ ($\beta_2 \geq \beta_1 \geq 1$) Pham (1991) proposed

$$K_R = \frac{AR}{V} \frac{1}{1 + \beta_1^{-2} + \beta_2^{-2}} \quad (4.35)$$

where AR/V is calculated by Eq. 4.19 or 4.23. For $\beta_1 < 2.0$, $\beta_2 < 2.0$ the following approximation can be used with less than 2% error:

$$K_R = \frac{1 + \beta_1^{-1} + \beta_2^{-1}}{1 + \beta_1^{-2} + \beta_2^{-2}} \qquad (4.36)$$

For two-dimensional ellipses (infinitely long ellipsoidal cylinders) simply put $\beta_2^{-1} = 0$ in the above equations.

4.3.4   Arroyo & Mascheroni's Method

In this method, which could be interpreted as a combination of the shape factor and MCP approaches, the freezing time as calculated by Salvadori & Mascheroni's formula (Eq. 4.8) is modified by using two empirical factors (Arroyo and Mascheroni 1990; Salvadori and Masheroni, 1996):

$$t_f = K_1 \left(A\theta_c + B \right) \left(\frac{1}{Bi_u} + K_2 C \right) \left(1 - \frac{\theta_i}{\theta_f} \right)^n \left(\frac{\theta_a}{\theta_f} - 1 \right)^{-m} \frac{R^2}{\alpha_u} \qquad (4.37)$$

where the shape-dependent factors K_1, K_2 are listed in Table 4.8 and the other empirical factors are the same as for Eq. 4.8.

4.3.5   Comparison of Shape Factor and Mean Conducting Path Approaches

Assuming that the freezing time can be accurately calculated by Plank's equation (Eq. 3.14) or Pham's (1986) equation (Eq. 4.4), i.e. that the total freezing time is a sum of two terms, one due to internal resistance and the other to external resistance, then for a product with volume V and surface area A:

$$\frac{1}{E_f} \left(\frac{V}{A} \right)_{slab} \left(1 + \frac{Bi_f}{2} \right) = \frac{V}{A} \left(1 + K_R \frac{Bi_f}{2} \right) \qquad (4.38)$$

The left-hand side uses the shape factor approach, while the right hand side uses the MCP approach. Putting $(V/A)_{slab} = R$ where R is the minimum distance from surface to thermal centre, we then obtain the following relationships between shape factor and MCP:

$$E_f = \frac{AR}{V} \frac{1 + \frac{Bi_f}{2}}{1 + K_R \frac{Bi_f}{2}} \qquad (4.39)$$

Table 4.8 Freezing shape factors to be used in Eq. 4.37

Shape	K_1	K_2
Infinite rectangular rod	$\dfrac{1}{1+\beta_1^{-2}}$	1
Brick	$\dfrac{1}{1+\beta_1^{-2}+\beta_2^{-2}}$	$\left[1+\left(\dfrac{\beta_1}{\beta_2}\right)^2\right]^{1/2}$
Finite cylinder	$\dfrac{1}{2+\beta_1^{-2}}$	$\left(1+\beta_1^{-2}\right)^{1/2}$

$$K_R = \frac{1}{E_f}\frac{AR}{V}\left(1+\frac{2}{Bi_f}\right)-\frac{2}{Bi_f} \qquad (4.40)$$

(These equations were first presented by Cleland et al. (1987a); note however that the MCP in that work, as in some other papers from that period, is twice the MCP as defined in this book.)

Since E_f and K_R can be calculated from each other, there is no strong reason to prefer one approach over the other concerning prediction accuracy. However, for small to moderate values of the ratio of the longest and shortest dimensions, the MCP approach is preferable as the theoretical basis is easy to visualise. For squat objects (with aspect ratio close to 1) such as square rods, cubes, ellipsoids and finite cylinder with $\beta \approx 1$, and especially if $Bi \ll 1$, the geometric mean of R_{max} and R_{min} can be taken as the MCP. In other cases, the available formulas for E_f are probably better established and more reliable.

4.4 Thawing Time Prediction

Although freezing and thawing both involve phase change, empirical freezing time prediction formulas cannot be directly applied to thawing because they are not entirely symmetrical and process conditions usually differ. Because of microbiological concerns, the thawing medium temperature is usually quite low (around 5 °C) and close to the freezing point, while during freezing the difference is usually 20 K or more. Calorimetric properties are asymmetrical with respect to temperature, with a gradual phase change below the freezing point (Figs. 2.3–2.4), thus the freezing point is approached more gradually during thawing than during freezing.

During practical thawing situations, many complicating phenomena may occur. Water may drip out of the product, changing its properties. The product may soften and change its shape. If a block of food is made of distinct pieces these may fall out as the ice thaws, and in immersion freezing the process liquid may penetrate between the pieces. The calculation methods presented below should be used with caution if any of the above happens.

4.4.1   Cleland et al.'s method

Cleland et al. (1986) gathered extensive data on the thawing of various shapes
and presented four different equations which all agreed with data to within about
$\pm 10\%$. One of them, written in a Plank-like form, is as follows:

$$t_t = \frac{\rho R}{E_f h} \frac{\Delta H_{10}}{(T_a - T_f)} \left(P_1 + P_2 Bi_u \right) \tag{4.41}$$

where $Bi = hR/k_u$ and

$$P_1 = 0.7754 + 2.2828 Ste_{thaw} \cdot Pk_{thaw}$$

$$P_2 = 0.5 \left(0.4271 + 2.1220 Ste_{thaw} - 1.4847 Ste_{thaw}^2 \right)$$

$$Ste_{thaw} = \rho_u c_u \left(T_a - T_f \right) / \rho \Delta H_{10}$$

$$Pk_{thaw} = \rho_f c_f \left(T_f - T_i \right) / \rho \Delta H_{10}$$

Note that the freezing shape factor E_f is also used for thawing (1 for slab, 2 for
infinite cylinder, 3 for spheres). ΔH_{10} is the enthalpy change (J·kg^{-1}) of the product
from 0 to $-10\,^{\circ}\mathrm{C}$.

Ilicali et al. (1999) suggested the following alternative expressions for the thawing shape factor:

- for infinite cylinders:

$$E_f = 2.0422 + 0.0857\theta_c - 0.00076\theta_i - 0.00356\theta_a + \frac{0.0099}{2Bi_u} \tag{4.42}$$

- for spheres:

$$E_f = 3.127 + 0.1857\theta_c - 0.00168\theta_i - 0.00818\theta_a + \frac{0.02034}{2Bi_u} \tag{4.43}$$

4.4.2   Salvadori and Masheroni's Method

Salvadori and Masheroni (1991) proposed the following correlation, which has the
same form as their freezing time equation:

Table 4.9 Parameter values and variable ranges for Salvadori & Mascheroni's method (Eqs. 4.44 and 4.45)

Geometry	A	B	C	m	n	A'	B'	C'	m'	n'
Slab	0.3500	26.40	0.45	0.740	0.03	0.321	23.637	0.435	0.763	0.099
Infinite cylinder	0.1685	12.13	0.47	0.740	0.05	0.109	12.572	0.428	0.707	0.032
Sphere	0.0100	8.46	0.45	0.715	0.03	0.039	8.120	0.408	0.671	0.027

$$t_t = \left(A\theta_c + B \right) \left(\frac{1}{Bi_u} + C \right) \left(1 - \frac{\theta_i}{\theta_f} \right)^n \left(\frac{\theta_a}{\theta_f} - 1 \right)^{-m} \frac{R^2}{\alpha_u} \qquad (4.44)$$

Later (Salvadori 1994) they modified the equation to

$$t_t = \left(A'\theta_c + B' \right) \left(\frac{1}{Bi_u} + C' \right) \left(-1 - \theta_i \right)^{n'} \left(\theta_a + 1 \right)^{-m'} \frac{R^2}{\alpha_u} \qquad (4.45)$$

Values of the parameters are listed in Table 4.9.

4.4.3   Thawing Time Prediction for Multi-Dimensional Shapes

Hossain et al. (1992a, b, c) showed that the analytical shape factors derived for Plank's equation (Eqs. 3.20–3.25) can also be used for thawing. However, if the user is not be able to calculate infinite series, Cleland et al. (1987a, b) empirical expression can also be used (Eq. 4.15).

For ellipses Ilicali et al. (1999) suggested

$$E_f = E_{Pham} - 0.01407 + 0.03303\theta_c + \frac{0.01968}{\beta Bi} \qquad (4.46)$$

where E_{Pham} is the value calculated by Eq. 4.17.

4.5   Freezing Heat Load

4.5.1   Total and Average Heat Load

Knowledge of the product heat load is important for designing refrigeration systems. The total (cumulative) heat load (in J) from freezing a product can be easily calculated from

$$Q_{cum} = \rho V (H_i - H_e) \qquad (4.47)$$

Fig. 4.5 Temperatures
and dynamic heat load for
a rectangular box of food
during freezing (sides
0.15 m × 0.15 m × 0.15 m,
$T_i=30\,°C$, $T_a=-30\,°C$,
$h=10$ W·m^{-2}K^{-1}), calculated
by a rigorous numerical
method for a rectangular box.
Vertical dashed lines indicate
start and end of effective
"phase change period"

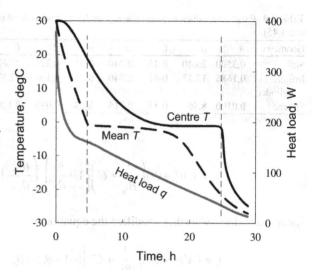

where H_i and H_e are the mean product enthalpy (J·kg^{-1}) at the beginning and end of
the process respectively, and can be calculated from specified product temperatures
using experimental calorimetric data or the methods described in Chap. 2. For continuous freezers with a throughput of N product units per unit time, the average heat
load Q_{av} (W) is given by

$$Q_{av} = NQ_{cum} = N\rho V\left(H_i - H_e\right) \tag{4.48}$$

For batch freezers, the heat load will decrease with time as the product cools and
the dynamic heat load must be calculated. Figure 4.5 shows a typical dynamic heat
load profile, calculated by a rigorous numerical method for a rectangular box of
food, together with centre and mean temperatures. The mean temperature is defined
as that corresponding to the mean enthalpy of the product according to the *H-T*
relationship. Even though the corners of the box would start to freeze within a very
short time, for the purpose of calculations we shall arbitrary define the phase change
period as starting when the mean temperature reaches freezing point and terminat-
ing when the centre temperature falls below it, as shown by the vertical dashed lines
in Fig. 4.5.

Some general features could be distinguished from the graph. During the precooling period (before the left vertical line), the heat load falls from its initial value
very quickly, as the corners, edges and surface layers of food rapidly lose heat, then
more gradually. The initial heat load Q_{init} (W) can be readily calculated:

$$Q_{init} = hA\left(T_i - T_a\right) \tag{4.49}$$

During the "phase change period", the heat load falls almost linearly as latent heat
is released and this pattern continues into the subcooling period.

From the practical design point of view, the subcooling heat load is negligible and therefore, the subcooling period can be lumped together with the phase change period, as assumed in Pham's (1986) freezing time equation (Eq. 4.4). The initial heat load peak (Eq. 4.49), which only lasts less than 1 % of the process duration, is also of little importance: in practice, that heat will be absorbed by the cold surroundings (walls, ceiling, supporting structures etc.) which will then release it slowly back into the cooling system, effectively smearing out the peak. It would be therefore very uneconomical to try to design to the initial peak load. Instead, we can assume a more gradual dynamic heat load for the precooling period, which stars well below the actual initial value given by Eq. 4.49.

The approach we follow in this chapter is from Lovatt et al. (1993a, b) with a few simplifications in view of the above considerations. Approximate dynamic heat load calculations of this type are used in the simulation of industrial plants, e.g. in the RADS package (Cleland 1985, 1990; Lovatt 1998) or MIRINZ Food Product Modeller (Agresearch 2013), in which many different products are cooled or frozen under different conditions. The rigorous methods of the next chapter will give more accurate results, but are too time consuming for this type of simulation.

In summary, we divide the heat load curve into a precooling period and a phase change-subcooling period. We will consider the second period first, for reasons to be explained later.

4.5.2 Heat Load During the Phase Change-Subcooling Period

By assuming that the product is divided into a frozen zone with volume V_f and an unfrozen zone with volume $V - V_f$, with enthalpies differing by $\Delta H_2 =$ latent heat + subcooling heat (defined as in Eq. 4.4), the heat load from freezing and subcooling becomes:

$$Q_{freeze} = \rho \Delta H_2 \frac{dV_f}{dt} \qquad (4.50)$$

Assuming further that the unfrozen zone has a radius or half-dimension r_f (the freezing front) we can write

$$Q_{freeze} = \rho \Delta H_2 \frac{dr_f}{dt} \frac{dV_f}{dr_f} \qquad (4.51)$$

dr_f/dt is the freezing front velocity and, for a slab or a sphere, can be derived in a manner similar to that used in deriving Plank's equation (Lovatt et al. 1993a, b):

$$\frac{dr_f}{dt} = -\frac{T_{fm} - T_a}{\rho \Delta H_2 r_f^{E_f - 1} \left[\dfrac{1}{hR^{E_f - 1}} + \dfrac{R^{2-E_f} - r_f^{2-E_f}}{k_f(2 - E_f)} \right]} \qquad (4.52)$$

where T_{fm} is calculated from Eq. 4.5, with the initial condition

$$r_f = R \quad \text{at} \quad t = t_{precool} \tag{4.53}$$

where $t = t_{precool}$ is the duration of precooling, to be determined later. The reader might verify that for the slab ($E_f = 1$) Eq. 4.52 reduces to Eq. 4.4. dV_f/dr_f represents the change of frozen volume with frozen depth. It is a function of geometry and the following approximate relationship is assumed:

$$\frac{V - V_f}{V} = \left(\frac{r_f}{R}\right)^Z \tag{4.54}$$

where Z is a parameter that measures the dimensionality of the freezing front. Differentiating:

$$\frac{dV_f}{dr_f} = -\frac{ZV}{R}\left(\frac{r_f}{R}\right)^{Z-1} \tag{4.55}$$

For a slab, infinite cylinder or sphere, $Z = E_f = AR/V$, while for other shapes Z must be found by curve-fitting data or numerical predictions. If no previous data is available, use

$$Z = \min\left(\frac{AR}{V}, 3\right) \tag{4.56}$$

Substituting Eqs. 4.52 and 4.55 into Eq. 4.51 and re-arranging gives

$$Q_{freeze} = \frac{s^{Z-E_f}}{1 + \frac{1 - s^{2-E_f}}{2 - E_f} Bi_f} \frac{ZV}{R} h\left(T_{fm} - T_a\right) \tag{4.57}$$

where

$$s = \frac{r_f}{R} \tag{4.58}$$

For slabs and spheres, putting $Z = E_f = AR/V$ leads to

$$Q_{freeze} = \frac{hA\left(T_{fm} - T_a\right)}{1 + \frac{1 - s^{2-E_f}}{2 - E_f} Bi_f} \tag{4.59}$$

This equation has a singularity at $E_f = 2$, so for infinite cylinder a value close to 2 (Lovatt et al. used 2.01) has to be used as an approximation. Alternatively, we can use the average of heat loads calculated with $E_f + \delta$ and $E_f - \delta$, where δ is a small number.

If environmental conditions (T_a and h) are constant, Eq. 4.52 can be integrated to give an explicit equation for t vs. r_f (Lovatt et al. 1993a).

4.5.3 Dynamic Heat Load During the Precooling Period

4.5.3.1 Analytical Solutions

Solution of the heat conduction equation in one dimension during chilling (no phase change) leads to the solution (Carslaw and Jaeger 1959)

$$T_m = T_a + (T_i - T_a) \sum_{i=1}^{\infty} a_i e^{-b_i t} \tag{4.60}$$

where the coefficients a_i, b_i depend on geometry and Biot number. From the mean temperature T_m, the enthalpy of the product and hence its heat content can be calculated, and the dynamic heat load $Q(t)$ can be calculated from

$$Q(t) = -\rho c V \frac{dT_m}{dt} = \rho c V (T_i - T_a) \sum_{i=1}^{\infty} a_i c_i e^{-b_i t} \tag{4.61}$$

After about 30 % of the initial temperature difference $T_i - T_a$ has dissipated, only the first term of the infinite series remains significant, in what we call the exponential decay period. The mean temperature then varies according to

$$T_m = T_a + (T_i - T_a) a_1 e^{-b_1 t} \tag{4.62}$$

Equations and graphs for the coefficients a_1, b_1 for simple geometries are available from many sources (e.g. Carslaw and Jaeger 1959; Pflug et al. 1965). Simple approximate equations were proposed by Ramaswamy et al. (1982) for simple shapes. For multidimensional geometries approximate equations were proposed by Smith et al. (1968), Hayakawa and Villalobos (1989) and Lin et al. (1993, 1996a, b).

For practical freezing processes most or all of the precooling may fall into the initial 30 % period during which the exponential decay approximation is not accurate, so these equations are not very useful.

4.5.3.2 Lovatt et al.'s Method

Lovatt et al (1993a) recommended the following approximate equation, based on Cleland and Earle (1982):

$$\frac{dT_m}{dt} = -\frac{E_c}{3} \frac{\phi^2 k_u}{\rho c_u R^2} (T_m - T_a) \tag{4.63}$$

$$Q_{precool} = -\rho c_u V \frac{dT_m}{dt} = \frac{E_c}{3} \frac{V}{R^2} \phi^2 k_u (T_m - T_a) \tag{4.64}$$

with the initial condition

$$T_m = T_i \quad \text{at} \quad t = 0 \tag{4.65}$$

where E_c is the shape factor (similar to E_f) for cooling without phase change, and ϕ is the first root of

$$\phi \cot \phi + Bi_u - 1 = 0 \qquad (4.66)$$

Lovatt et al. assumed $E_c = E_f$, but specific expressions for the cooling shape factor have been presented in more recent work (Lin et al. 1993, 1996a, b; Bart and Hanjalic 2003). They also recommend that Eq. 4.64 be applied until the calculated heat load is equal to or less than that calculated by the formula for the phase change period (given in Sect. 4.5.2), at which point the latter is used.

4.5.3.3 Pham's Method

The simplified method proposed here is based on Pham's (1986) freezing time equation, Eq. 4.4. The cooling rate is given by Newton's cooling law with a factor $(1 + Bi/2)$ to correct for internal resistance:

$$Q_{precool} = \frac{hA(T_m - T_a)}{1 + \dfrac{Bi_f}{2}} \qquad (4.67)$$

The precooling period is considered to end when $T_m = T_f$ (not T_{fm}). To avoid an upward jump in the heat load at the transition point, we must not let the precooling heat load fall below the heat load at the beginning of the phase change-subcooling period. The latter can be found by putting $s = 1$ in Eq. 4.57:

$$Q_{freeze}(s = 1) = \frac{ZV}{R} h(T_{fm} - T_a) \qquad (4.68)$$

hence

$$Q_{precool} = \max \left[\frac{hA(T_m - T_a)}{1 + \dfrac{Bi_f}{2}}, \frac{ZV}{R} h(T_{fm} - T_a) \right] \qquad (4.69)$$

The mean temperature is given by

$$\frac{dT_m}{dt} = \frac{-Q_{precool}}{\rho c_u V} = \frac{-1}{\rho c_u V} \max \left[\frac{hA(T_m - T_a)}{1 + \dfrac{Bi_f}{2}}, \frac{ZV}{R} h(T_{fm} - T_a) \right] \qquad (4.70)$$

with the initial condition

$$T_m = T_i \quad \text{at} \quad t = 0 \qquad (4.71)$$

If environmental conditions (T_a and h) are constant, T_m and hence $Q_{precool}$ can be explicitly expressed:

$$T_m = T_a + (T_i - T_a) e^{-at} \qquad (4.72)$$

$$Q_{precool} = \frac{hA(T_i - T_a)}{1 + \dfrac{Bi_f}{2}} e^{-at} \qquad (4.73)$$

where

$$a = \frac{-hA}{\rho c_u V \left(1 + \dfrac{Bi_f}{2}\right)} \qquad (4.74)$$

4.5.4 Summary of Method

The dynamic heat load determination method of this chapter consists of the following steps:

1. Determine the shape factors E_f and Z. If a "best fit" value of Z is not known, use Eq. 4.56.
2. Integrate Eq. 4.70 to calculate the mean temperature T_m at various times, until $T_m \leq T_f$.
3. Calculate the precooling heat load $Q_{precool}$ at various times from Eq. 4.69.
4. After T_m reaches T_f, integrate Eq. 4.52 to calculate the freezing front position r_f at various time, until $r_f = 0$.
5. Calculate the phase change-subcooling heat load Q_{freeze} at various times from Eq. 4.57.

Figure 4.6 shows a comparison of the dynamic heat load calculated by the above method for the freezing of a carton of lean meat (0.15 m × 0.30 m × 0.60 m sides, cooled from 30 °C in environment at −30 °C and $h = 10$ W·m^{-2}K^{-1}) with that calculated by a rigorous numerical method (finite volumes). For the phase change-subcooling period two values of Z were tried, $Z = AR/V = 1.75$ and $Z = 1.95$, the latter giving much better fit. The very high initial peak is smeared out due to the exponential decay approximation for precooling, and the heat load falls off sharply at the end of freezing due to our lumping of the phase change and subcooling period. Both these discrepancies are probably of little practical significance.

4.6 Summary and Recommendations

- The recommended simple (non-numerical) methods for freezing time prediction are Pham's (1984) method (Eq. 4.3) and Pham's (1986) method (Eq. 4.4). Both are based on Plank's equation with additional terms for the pre- and post-cooling times. Pham's first method is less empirical and therefore probably more robust when unusual conditions are encountered, Pham's second method is simpler to apply.

Fig. 4.6 Heat load for the freezing of a meat carton by a rigorous method (finite volumes) and by the approximate method of this chapter

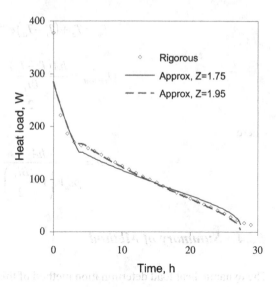

- When the properties of the frozen food are lacking, Salvadori & Mascheroni's method (Eq. 4.6) may be used. However, at initial temperatures around $0\,^{\circ}\text{C}$ this formula predicts unrealistic effects of initial temperature.
- For extended freezing conditions (foods with low moisture and/or low freezing point, cryogenic freezing temperatures) correction factors have been derived for Pham's two methods (Eqs. 4.11 and 4.12) that produce results in agreement with numerical calculations to within about 10 %.
- Empirical shape factors E_f for ellipses, ellipsoids and irregular shaped foods are given. For other regular geometries the analytical shape factors given in Chap. 3 may be used.
- For products with small to moderate values of the aspect ratio (ratio of longest to shortest dimension), the mean conducting path MCP is a simple alternative approach to the shape factor E_f. When the aspect ratio is close to 1, and especially if $Bi \ll 1$, the geometric mean of R_{max} and R_{min} can be taken as the MCP, which can then be used in calculating the effective Biot number.
- Empirical formulas are presented for thawing time. However, if the product breaks up or is penetrated by the thawing medium then the process time will be reduced.
- A method for predicting the dynamic heat load during food freezing is presented and summarised in Sect. 4.5.4.

 **CAUTION**

Older publications in this field often define the Biot number as $2hR/k$ instead of hR/k. Make sure that the correct interpretation is applied.

Chapter 5
Numerical Methods

5.1 Introduction

Rigorous numerical methods can provide precise and accurate solutions to any physical problem, as long as all the important physical phenomena involved are properly taken into account. They do this by discretizing space and time to convert the governing partial differential equations (PDEs), into a set of algebraic equations that are then solved. For the freezing of a solid governed purely by heat transfer, the governing equation is the heat conduction equation (introduced in Chap. 3):

$$\rho c \frac{\partial T}{\partial t} = \nabla \cdot (k \nabla T) + S_q \qquad (3.5)$$

Some calculation methods include the latent heat of phase change in the source term S_q, in which case c is the sensible specific heat, while other methods include the latent heat in the term on the left, in which case c is the apparent specific heat c_{app}. When considering numerical methods (Sects. 5.2 and 5.3) we shall use the symbol c to mean either of the above, depending on context.

The first step is to discretize the space domain by superimposing on it a grid consisting of sets of points called *nodes*. The continuous temperature field is represented by a set of *nodal temperatures,* represented by a vector $\mathbf{T} = (T_1, T_2, T_3 \ldots)$. The temperature at each node is influenced by those of the surrounding nodes. The PDE is then transformed into a system of ordinary differential equations (ODEs):

$$\mathbf{C} \frac{d\mathbf{T}}{dt} + \mathbf{K}\mathbf{T} = \mathbf{f} \qquad (5.1)$$

$\mathbf{C}$ is the *global capacitance matrix* containing the specific heat c (or c_{app}) and characterising the thermal inertia of the system,
$\mathbf{K}$ the global conductance matrix containing the thermal conductivity k and
$\mathbf{f}$ the global forcing vector containing known terms arising from heat generation and boundary conditions.

Q. T. Pham, *Food Freezing and Thawing Calculations,*
SpringerBriefs in Food, Health, and Nutrition, DOI 10.1007/978-1-4939-0557-7_5,
© The Author 2014

Both $\mathbf{C}$ and $\mathbf{K}$ are sparse matrices, where C_{ij} and K_{ij} are non-zero only if i and j denote neighbouring nodes. In the finite difference and finite volume methods, $\mathbf{C}$ is a diagonal matrix.

To solve for the temperature history, we must specify the initial and boundary conditions, the latter being usually one of three main types:

1. First type, or Dirichlet boundary condition, when the boundary temperature T_s is prescribed.
2. Second type, or Neumann boundary condition, when the heat flux q_s at a boundary S is prescribed. In one dimension, this is expressed as follows, where the temperature gradient is calculated on the product side of the surface:

$$q_s = -k\frac{\partial T}{\partial x}\bigg|_s \tag{5.2}$$

3. Third type, or Robin boundary condition, when the heat flux at the boundary is equal to $h(T_s - T_a)$ where the heat transfer coefficient h is prescribed. In one dimension:

$$h(T_a - T_s) = -k\frac{dT}{dx}\bigg|_s \tag{5.3}$$

Once the ODE system is obtained, it can be integrated by a variety of time-stepping methods to yield the nodal temperature array at various times.

5.2   Discretization of the Space Domain

The most popular numerical methods for discretizing the space domain are the finite difference method (FDM), the finite elements method (FEM) and the finite volume method (FVM). FDM requires a structured grid (one with a regular orthogonal array of nodes), and therefore, is normally used only for regular shapes in Cartesian, cylindrical or spherical coordinates: slab, cylinder, sphere, rectangular rod or brick. However, by the use of boundary-fitted coordinates, its use could be extended to some other shapes. FEM and FVM can be used with structured or unstructured grids, and are therefore, more useful for practical applications where the product's shape may be complex and irregular.

This chapter does not aim to be a tutorial on numerical methods. However, we will use a simple example to illustrate their basic principles, main features, similarities and differences, before examining their application to food freezing problems. For an infinite slab conducting heat in the x direction only, without internal heat generation, Eq. 3.5 becomes

$$\rho c\frac{\partial T}{\partial t} = \frac{\partial}{\partial x}\left(k\frac{\partial T}{\partial x}\right) \tag{5.4}$$

Fig. 5.1 Finite difference grid

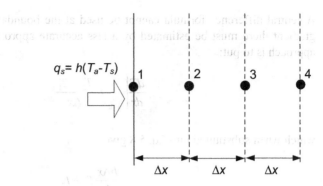

$$q_s = h(T_a - T_s)$$

We shall assume the third type b.c., Eq. 5.3, as it is the most frequent condition encountered in practice.

5.2.1   Finite Difference Method (FDM)

An FDM grid is obtained by placing nodes at the two surfaces and at equal distances in-between, as in Fig. 5.1. (The nodes do not have to be equally spaced, but this makes the formulas simpler.) Only a few nodes near the left wall are shown for simplicity.

The temperature gradient between nodes i and $i+1$ is found by central differences:

$$\left(\frac{\partial T}{\partial x}\right)_{i+} = \frac{T_{i+1} - T_i}{\Delta x} \tag{5.5}$$

Similarly, that between nodes $i-1$ and i is:

$$\left(\frac{\partial T}{\partial x}\right)_{i-} = \frac{T_i - T_{i-1}}{\Delta x} \tag{5.6}$$

Equation 5.4 then becomes, in the absence of heat sources:

$$\rho c \frac{\partial T_i}{\partial t} = \frac{k_+ (T_{i+1} - T_i) - k_- (T_i - T_{i-1})}{\Delta x^2} \tag{5.7}$$

where k_+ is the effective thermal conductivity between nodes i and $i+1$, and similarly for k_-. This equation has to be modified for the boundary node, $i=1$, since node 0 does not exist. The boundary condition of Eq. 5.3 becomes:

$$h(T_a - T_1) = -k \frac{\partial T}{\partial x}\bigg|_{x=+0} \tag{5.8}$$

A central difference formula cannot be used at the boundary, so the temperature gradient there must be estimated by a less accurate approximation method. One approach is to put:

$$\left.\frac{dT}{dx}\right|_{x=+0} = \frac{T_2 - T_1}{\Delta x} \tag{5.9}$$

which when substituted into Eq. 5.8 gives:

$$T_1 = \frac{\dfrac{h\Delta x}{k}T_a + T_2}{\dfrac{h\Delta x}{k} + 1} \tag{5.10}$$

This formula is only first-order accurate and approximates the gradient at $x = \Delta x/2$ rather than at $x = 0$. A better approximation is obtained by assuming a parabolic profile between the first three nodes, giving:

$$\left.\frac{dT}{dx}\right|_{x=+0} = \frac{-3T_1 + 4T_2 - T_3}{2\Delta x} \tag{5.11}$$

which when substituted into Eq. 5.8 gives:

$$T_1 = \frac{\dfrac{h\Delta x}{k}T_a + 2T_2 - 0.5T_3}{\dfrac{h\Delta x}{k} + 1.5} \tag{5.12}$$

Either approach will allow T_1 to be algebraically calculated from the other temperatures; hence, the corresponding ODE is eliminated from the system of equations. It can be easily verified that the same happens if the boundary condition is of the first or second type. Equation 5.7 can then be written for nodes 2, 3, 4 etc. in the form:

$$\begin{cases} C_{22}\dfrac{dT_2}{dt} = -K_{22}T_2 - K_{23}T_3 + f_2 \\[2mm] C_{33}\dfrac{dT_3}{dt} = -K_{32}T_2 - K_{33}T_3 - K_{34}T_4 + f_3 \\[2mm] C_{44}\dfrac{dT_4}{dt} = -K_{43}T_3 - K_{44}T_4 - K_{45}T_5 + f_4 \\[2mm] \qquad\qquad \cdots \end{cases}$$

Fig. 5.2 Error in the
temperature profile near the
surface of a finite difference
grid on applying a surface
temperature change

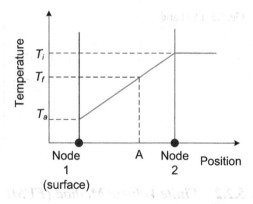

or, in matrix form:

$$
\begin{bmatrix} C_{22} & 0 & 0 & \dots \\ 0 & C_{33} & 0 & \dots \\ 0 & 0 & C_{44} & \dots \\ \dots & \dots & \dots & \dots \end{bmatrix} \begin{bmatrix} \dot{T}_2 \\ \dot{T}_3 \\ \dot{T}_4 \\ \dots \end{bmatrix} - \begin{bmatrix} K_{22} & K_{23} & 0 & \dots \\ K_{32} & K_{33} & K_{34} & \dots \\ 0 & K_{43} & K_{44} & \dots \\ \dots & \dots & \dots & \dots \end{bmatrix} \begin{bmatrix} T_2 \\ T_3 \\ T_4 \\ \dots \end{bmatrix} = \begin{bmatrix} f_2 \\ f_3 \\ f_4 \\ \dots \end{bmatrix} \quad (5.13)
$$

which is the expanded form of Eq. 5.1. The forcing term f_i contains known quantities from the boundary conditions, and is zero for nodes that are not near the boundaries. It may be noticed that the capacitance matrix $\mathbf{C}$ is diagonal and the conductance matrix $\mathbf{K}$ is tridiagonal, each line of the equation containing only the temperatures of three immediate neighbouring nodes. These features lead to some easy and fast methods for solution, such as the explicit Euler time-stepping method, where the nodal equations are solved one by one, or the solution of the ODEs by the tridiagonal matrix algorithm.

FDM can easily handle the first type of boundary conditions, but with the second and third type the approximation involved in Eq. 5.9 or 5.11 is not very satisfactory unless the grid is very fine, due to the single-sided estimate of the temperature gradient at the boundary. In all cases, heat conservation is not ensured (Patankar 1980; Botte et al. 2000). This is particularly serious for phase change problems. For example, on applying a prescribed surface temperature condition, the surface node immediately falls to the prescribed temperature T_a which may be below the freezing point, giving the nodal temperature profile in Fig. 5.2 (if a linear profile is assumed), and implying that the freezing front immediately jumps to point A, while the front would actually be still at the surface at that time. The latent heat contained in the layer between the surface and depth A will instantaneously vanish and a significant heat balance error will result. For this reason, a control volume formulation (finite volume method) is preferred.

Fig. 5.3 FVM grid

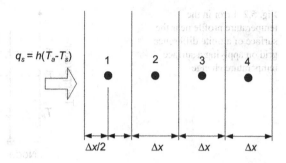

$$q_s = h(T_a - T_s)$$

5.2.2   Finite Volume Method (FVM)

While the terminology is relatively new, FVM in some primitive form has been used by engineers for a long time (even before the days of computers) in view of its conceptual clarity. Some textbooks derive the FVM method by integrating the PDE governing heat or mass transport over a control volume, but this is not really necessary, at least for orthogonal grids, since the PDEs were derived by taking the infinitesimal limit of a control volume in the first place (Bird et al. 1960).

A FVM grid can be obtained by dividing the domain into control volumes and placing a node at the centre of each volume, as in Fig. 5.3. For simplicity, equal control volumes are used here but this is not necessary. The temperature at each node represents the mean temperature of the surrounding control volume.

The nodal arrangement is similar to FDM except for the boundary grids, which are not situated at the surface. Per unit area, the volume of each control volume is Δx. The heat fluxes between neighbouring nodes remain as given in Eqs. 5.5 and 5.6. An energy balance over each control volume gives:

$$\rho c \Delta x \frac{\partial T_i}{\partial t} = k_- \frac{T_{i-1} - T_i}{\Delta x} + k_+ \frac{T_{i+1} - T_i}{\Delta x} \tag{5.14}$$

which is identical to Eq. 5.7 for FDM. The difference is in the treatment of the boundary condition. By addition of thermal resistances, the heat flux from the surroundings to Node 1 is given by:

$$q_s = \frac{1}{\frac{1}{h} + \frac{\Delta x}{2k}} (T_a - T_1) \tag{5.15}$$

A heat balance over the control volume around Node 1 then leads to:

$$\rho c \Delta x \frac{\partial T_i}{\partial t} = \frac{1}{\frac{1}{h} + \frac{\Delta x}{2k}} (T_a - T_1) - k \frac{T_1 - T_2}{\Delta x} \tag{5.16}$$

or

$$\rho c \frac{\partial T_i}{\partial t} = -\left(\frac{1}{\frac{\Delta x}{h} + \frac{\Delta x^2}{2k}} + \frac{k}{\Delta x^2} \right) T_1 + \frac{k}{\Delta x^2} T_2 + \frac{T_a}{\frac{\Delta x}{h} + \frac{\Delta x^2}{2k}} \tag{5.17}$$

which could be written as:

$$C_{11} \frac{dT_1}{dt} = -K_{11}T_1 - K_{12}T_2 + f_1 \tag{5.18}$$

This results in a matrix ODE similar to Eq. 5.13, except that the nodal temperature array includes node T_1. Because central differences are used throughout the domain for calculating the heat fluxes, FVM is more accurate than FDM and energy balance is ensured (Botte 2000). The second type boundary condition is handled by replacing the first term on the right of Eq. 5.16 by the prescribed value of q_s, while the first type b.c. is handled by putting $1/h = 0$.

The 1-D formulation above can be extended to infinite cylinders and spheres (including hollow ones), since in these geometries the heat equation can be written:

$$\rho c \frac{\partial T}{\partial t} = \frac{1}{x^n} \frac{\partial}{\partial x} \left(kx^n \frac{\partial T}{\partial x} \right) \tag{5.19}$$

where $n = 1$ for cylinders and 2 for spheres. In FVM, the nodes can be visualised as being leek- or onion-like layers. The heat balance (Eq. 5.14) has to be modified to take into account the different heat flow area on either side of each node.

5.2.3 Finite Element Method (FEM)

In FEM, the domain is divided into elements, which superficially resemble control volumes in FVM. However, instead of a single node at the centre, nodes are placed at each element's apexes (and also elsewhere in non-linear formulations). Within each element, the temperature field at x is approximated by interpolating between the nodes. Thus, for the 1-D example being considered, using linear interpolation:

$$T = (1 - \xi)T_A + \xi T_B \tag{5.20}$$

where $\xi = (x - x_A)/(x_B - x_A)$ is the relative distance between A and B (Fig. 5.4). This can be written in a more general way:

$$T(\mathbf{x}, t) = \mathbf{N}(\mathbf{x}) \cdot \mathbf{T}(t) \tag{5.21}$$

Fig. 5.4 Using shape functions N_A, N_B to interpolate temperature in an element AB

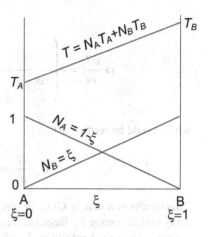

where $\mathbf{T}(t)$ are the vector of nodal temperatures and $\mathbf{N(x)}$ is a vector of position-dependent *shape functions*. In the present example the shape functions are:

$$\mathbf{N} = \begin{pmatrix} 1 - \xi \\ \xi \end{pmatrix} \tag{5.22}$$

Higher-order shape functions may also be used. However, in all cases N_i must be 1 at node i itself, and $\sum N_i = 1$ everywhere in the element.

Because the temperature field is only approximate, the heat conduction equation will in general not be satisfied exactly at every point in the element, but it is reasonable to require that energy be conserved over the element as a whole. This could be done by integrating the residual $\rho c \partial T / \partial t - \nabla(k \nabla T)$ over the whole element and setting it to zero, but that would give only one equation, which is not enough to solve for all the nodal temperatures. We need as many equations as there are nodes. This is obtained by requiring that the integrated weighted residuals should be zero when the residual is weighted towards each node by some function which is maximal at the node and decreases gradually with distance. In the Galerkin version of FEM, the shape functions, which have this property, are also used as weighting functions:

$$\begin{cases} \displaystyle\int_\Omega N_A \left[\rho c \frac{\partial T}{\partial t} - \nabla(k \nabla T) \right] d\Omega = 0 \\[2ex] \displaystyle\int_\Omega N_B \left[\rho c \frac{\partial T}{\partial t} - \nabla(k \nabla T) \right] d\Omega = 0 \end{cases}$$

or

$$\int_\Omega \mathbf{N} \left[\rho c \frac{\partial T}{\partial t} - \nabla(k \nabla T) \right] d\Omega = 0 \tag{5.23}$$

where integration is over the element domain, Ω. Since $\sum N_i = 1$, by summing the terms of the above vector equation it will be seen that if Eq. 5.23 is obeyed, energy will be conserved over the element as a whole as well. Substituting for T from Eq. 5.21 into Eq. 5.23 and solving yields a relationship between the nodal temperatures of the element of the following form:

$$\mathbf{C}^{el} \frac{d\mathbf{T}}{dt} + \mathbf{K}^{el}\mathbf{T} = \mathbf{f}^{el} \tag{5.24}$$

where

$$\mathbf{C}^{el} = \int_{\Omega} \rho c \mathbf{N}\mathbf{N}^{T} d\Omega \tag{5.25}$$

and (for a convective boundary condition):

$$\mathbf{f}^{el} = \int_{S} h T_a \mathbf{N} dS \tag{5.26}$$

$$\mathbf{K}^{el} = \int_{\Omega} k \mathbf{B}\mathbf{B}^{T} d\Omega + \int_{S} h \mathbf{N}\mathbf{N}^{T} dS \tag{5.27}$$

where

$$\mathbf{B} = \nabla^{T}\mathbf{N} = \left[\frac{\partial \mathbf{N}}{\partial x}, \frac{\partial \mathbf{N}}{\partial y}, \frac{\partial \mathbf{N}}{\partial z} \right] \tag{5.28}$$

Ω is the element domain and S its boundary. The superscript "*el*" indicates that this is a relationship between the nodes derived from one element only. For example, with a linear 1-D element of length Δx, assuming an internal element ($\mathbf{f}=0$) with constant properties, substituting for $\mathbf{N}$ from Eq. 5.22 into Eqs. 5.25–5.28 gives:

$$\rho c \Delta x \begin{bmatrix} \frac{1}{3} & \frac{1}{6} \\ \frac{1}{6} & \frac{1}{3} \end{bmatrix} \frac{\partial}{\partial t}\begin{pmatrix} T_A \\ T_B \end{pmatrix} + \frac{k}{\Delta x}\begin{bmatrix} 1 & -1 \\ -1 & 1 \end{bmatrix}\begin{pmatrix} T_A \\ T_B \end{pmatrix} = 0 \tag{5.29}$$

or

$$\begin{cases} \rho c \Delta x \left(\dfrac{1}{3}\dfrac{\partial T_A}{\partial t} + \dfrac{1}{6}\dfrac{\partial T_B}{\partial t} \right) = k\dfrac{T_A - T_B}{\Delta x} \\[2mm] \rho c \Delta x \left(\dfrac{1}{6}\dfrac{\partial T_A}{\partial t} + \dfrac{1}{3}\dfrac{\partial T_B}{\partial t} \right) = -k\dfrac{T_A - T_B}{\Delta x} \end{cases} \tag{5.30}$$

Details of how to obtain the matrices for more general situations can be found in standard texts on FEM such as Segerlind (1984) or Zienkiewicz (1991). Once Eq. 5.24 has been obtained for all the elements, they are added to give the global equation, Eq. 5.1. Most terms C_{ij} and K_{ij} will be zero, and only when two nodes i and j are connected by one or more element will these terms be non-zero.

While the conduction terms on the right of Eq. 5.30 are similar to those found in FDM and FVM (e.g. Eq. 5.14), the capacitance terms on the left are less intuitively obvious and arise from the fact that in FEM, heat capacity is distributed over the element rather than concentrated at the nodes. A rough physical interpretation will help understand how the method works. Each term of the $\mathbf{K}^{el}$ matrix, K_{ij}^{el}, represents the conductance between nodes i and j of the material within the element. In more than one dimensions, two or more elements may share two nodes, meaning that there are parallel conduction paths, and the conductances from all these elements must be added together. Within an element, the thermal energy $\rho c T$ at each point is in some sense attributed to the nodes according to the shape function, i.e. more towards the nearest node and less towards the farthest. When the temperature at a node i is changed, this affects the temperature profile and hence the thermal energy throughout the element, and therefore, C_{ij} represents the effect of a change in T_i on the thermal energy attributed to node j.

In the *lumped mass* or *diagonal mass matrix* version of FEM, all the thermal energy change due to a change in T_i is attributed to node i; hence, $\mathbf{C}$ becomes a diagonal matrix. For example, Eq. 5.29 for a linear 1-D element becomes:

$$\frac{\rho c \Delta x}{2}\begin{bmatrix} 1 & 0 \\ 0 & 1 \end{bmatrix}\frac{\partial}{\partial t}\begin{pmatrix} T_A \\ T_B \end{pmatrix} + \frac{k}{\Delta x}\begin{bmatrix} 1 & -1 \\ -1 & 1 \end{bmatrix}\begin{pmatrix} T_A \\ T_B \end{pmatrix} = 0 \qquad (5.31)$$

or

$$\begin{cases} \dfrac{\rho c \Delta x}{2}\dfrac{\partial T_A}{\partial t} = k\dfrac{T_B - T_A}{\Delta x} \\[2ex] \dfrac{\rho c \Delta x}{2}\dfrac{\partial T_B}{\partial t} = k\dfrac{T_A - T_B}{\Delta x} \end{cases} \qquad (5.32)$$

A similar equation pair can be obtained for the adjacent element BC:

$$\begin{cases} \dfrac{\rho c \Delta x}{2}\dfrac{\partial T_B}{\partial t} = k\dfrac{T_C - T_B}{\Delta x} \\[2ex] \dfrac{\rho c \Delta x}{2}\dfrac{\partial T_C}{\partial t} = k\dfrac{T_B - T_C}{\Delta x} \end{cases} \qquad (5.33)$$

Adding the second line of Eq. 5.32 to the first of Eq. 5.33 gives

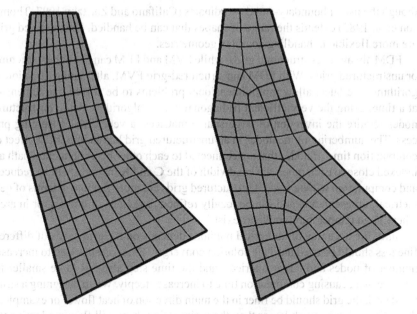

Fig. 5.5 Examples of structured (*left*) and unstructured (*right*) grids

$$\rho c \Delta x \frac{\partial T_B}{\partial t} = k \frac{T_A - T_B}{\Delta x} + k \frac{T_C - T_B}{\Delta x} \qquad (5.34)$$

which is identical to the FDM-FVM formulation, Eq. 5.14, although the resemblance is for the linear element only. In other words, the mass of each element is assumed to be concentrated at the nodes instead of being distributed over the element. This formulation has some advantages over the Galerkin formulation in terms of simplicity and stability (Banaszek 1989) and is particularly useful for dealing with the latent heat peak during freezing, as will be seen later. Lumped masses should be used with linear elements only, as higher-order elements will lead to negative nodal masses, which may adversely affect the accuracy and stability of the results.

5.2.4 Discretization of the Space Domain in 2-D and 3-D

Two- and three-dimensional objects can be discretized with structured (mapped) grids or unstructured (free) grids (Fig. 5.5). With structured grids, the nodes are systematically numbered in each dimension, so nodal temperatures are usually notated as a 2- or 3-D matrix, $[T_{i,j}]$ or $[T_{i,j,k}]$ (topological notation). The advantage of this notation is that neighbouring nodes can be immediately identified from the indices, so conducted fluxes can be easily calculated. Special solution techniques exist for structured grids which are very efficient in computing time. Structured grids are most suitable for regular shapes (rectangular rods or blocks, finite cylinders), although the use of boundary-fitted coordinates (Califano and Zaritzky 1997, Thompson et al. 1982) extends the range of shapes that can be handled. Unstructured grids are more flexible in handling complex geometries.

FDM always uses structured grids, while FVM and FEM can use either structured or unstructured grids. With FDM and structured-grid FVM, alternating direction algorithms (see later) allow multidimensional problems to be solved one dimension at a time, using the very efficient tridiagonal matrix algorithm, while unstructured nodes require the inversion of large sparse matrices, a very time-consuming process. The numbering of the nodes in an unstructured grid has an important effect on computation time. If nodes that are connected to each other by a conduction path are indexed close to each other, the bandwidth of the **C** and **K** matrices will be reduced, and computation will be faster. Unstructured grids are more flexible in terms of geometrical representation and can be locally refined in regions of interest, or in areas where large temperature gradients exist.

Since the accuracy of numerical methods depends on the grid, grids of different fineness should be tried until the solution converges. Finer meshes lead to increased number of nodes and larger matrices, and the time step also has to be smaller for convergence, causing computation time to increase steeply. When designing a structured grid, the grid should be finer in the main direction of heat flow. For example, if the x dimension is much larger than the y dimension, heat will flow predominantly in the y direction, and there should be more nodes in the y direction than in the x direction. Nodes don't have to be equally spaced, but changes in spacing should be as gradual as possible. Unstructured grids must be finer where thermal gradients are steep or complex, such as at corners, and the shape of the elements should not be overly skewed or elongated. Commercial FEM software nowadays can often ensure these requirements by automatic grid adaptation.

5.3   Time-Stepping

5.3.1   Two-Level Stepping Schemes

Having obtained a set of ODEs in time (Eq. 5.1) relating the nodal temperatures by discretizing the space domain, solution will proceed in a series of time steps starting from the known initial conditions. Several solution methods for ODEs are available (Scheerlink et al. 2001) such as Runge–Kutta methods, backward differentiation formulas or Crank–Nicolson (central differences). In heat conduction, the most popular class of methods involves the use of a time-averaged value of the nodal temperatures to calculate heat fluxes:

$$\mathbf{C}\frac{\mathbf{T}^{New}-\mathbf{T}}{\Delta t}=-\mathbf{K}\overline{\mathbf{T}}+\mathbf{f} \qquad (5.35)$$

with

$$\overline{\mathbf{T}} = \alpha \mathbf{T}^{New} + (1-\alpha)\mathbf{T} \tag{5.36}$$

where the superscript *New* refers to temperatures at the end of the time step (to be calculated). α is a parameter varying between 0 and 1. $\alpha = 0$ gives the Euler method for time-stepping, $\alpha = 1$ gives the backward stepping or fully implicit method and $\alpha = 0.5$ the Crank–Nicolson method, perhaps the most popular as it gives best accuracy combined with unconditional stability.

The Euler method can be written as

$$\mathbf{CT}^{New} = \mathbf{T} - \Delta t(\mathbf{KT} + \mathbf{f}) \tag{5.37}$$

In FDM, FVM and lumped mass FEM, **C** is diagonal and the above equation can be solved node by node to obtain $\mathbf{T}^{New}$:

$$T_i^{New} = T_i - \Delta t \left(\sum_j K_{ij} T_j + f_i \right) \tag{5.38}$$

For example, for an internal node in 1-D (Fig. 5.1) this becomes:

$$T_i^{New} = T_i + \frac{\Delta t}{\rho c \Delta x^2} \left[k_- (T_{i-1} - T_i) + k_+ (T_{i+1} - T_i) \right] \tag{5.39}$$

The Euler method is not used in consistent (distributed mass) FEM because the **C** matrix is not diagonal and it still has to be inverted.

Although easy to implement (and therefore, recommended for one-off programs or small problems), the Euler method is only first-order accurate and is unstable for $\frac{k\Delta t}{\rho c \Delta x^2} > \frac{1}{2}$. In two and three dimensions, the restriction is more stringent still: the upper limit becomes 1/4 and 1/6, respectively. Even when stable, the Euler method rapidly loses accuracy as Δt increases. The critical time interval decrease as the square of the space interval Δx; hence, the solution is very time-consuming when the domain is finely meshed.

For $\alpha = 1$ (backward difference), the new (and yet unknown) temperatures are used to compute the gradients:

$$\mathbf{CT}^{New} = \mathbf{T}^{New} - \Delta t(\mathbf{KT}^{New} + \mathbf{f}^{New}) \tag{5.40}$$

For an internal node in a 1-D lumped mass formulation (e.g. Fig. 5.1) this becomes

$$\rho c \Delta x \left(T_i^{New} - T_i \right) = \Delta t \left(k_- \frac{T_{i-1}^{New} - T_i^{New}}{\Delta x} + k_+ \frac{T_{i+1}^{New} - T_i^{New}}{\Delta x} \right) \tag{5.41}$$

or

$$-\frac{k_-\Delta t}{\Delta x}T_{i-1}^{New} + \left(\rho c\Delta x + \frac{k_+}{\Delta x} + \frac{k_-}{\Delta x}\right)T_i^{New} - \frac{k_+\Delta t}{\Delta x}T_{i+1}^{New} = \rho c\Delta x T_i \qquad (5.42)$$

The above equation is a linear relationship between three successive unknown nodal temperatures with indices $i-1$, i and $i+1$. These nodal equations can, therefore, be assembled into a matrix equation of the form:

$$\mathbf{AT}^{New} = \mathbf{B} \qquad (5.43)$$

where (in 1-D) $\mathbf{A}$ is a tridiagonal matrix, with all terms equal to zero except those on the diagonal, subdiagonal and superdiagonal, and $\mathbf{B}$ is a vector of known terms. The equation can be solved very efficiently by a simplified Gaussian elimination method (the Tridiagonal Matrix Algorithm or TDMA).

For $\alpha = 0.5$ (central difference or Crank–Nicolson), the arithmetic mean of old and new temperatures is used (Crank and Nicolson 1947), which also yields a tridiagonal matrix equation. The last mentioned is the most popular method in view of its combination of unconditional stability and second-order accuracy. Even then, the Crank–Nicolson scheme will be subject to large, slow-decaying oscillations when $\frac{k\Delta t}{\rho c\Delta x^2}$ is too large. For all values of $\alpha \neq 0$, unknown nodal temperatures appear on both sides of Eq. 5.35, and this set of equations has to be re-arranged to bring all the unknown temperatures to the left hand side.

5.3.2   Lee's Three-Level Scheme

Another popular time-stepping procedure in the food freezing literature is Lee's (1976) three-level scheme: temperature gradients are calculated from the mean of the temperatures at the present time, the previous time step, and the next time step. The resulting heat accumulation is used to calculate the temperature change between the *previous* time step and the next time step (instead of between the present time and the next time step):

$$\mathbf{C}\frac{\mathbf{T}^{New} - \mathbf{T}^{Old}}{\Delta t} = -\mathbf{K}\frac{\mathbf{T}^{Old} + \mathbf{T} + \mathbf{T}^{New}}{3} + \mathbf{f} \qquad (5.44)$$

In theory, this will allow rapid property changes to be better accounted for, however the author has not found any advantage of this scheme over the Crank–Nicolson scheme (Pham 1987c).

5.3.3 Use of Generic ODE Solvers

Commercial finite element and finite volume software such as Matlab, Ansys and Comsol use generic ODE solvers which are particularly suitable for highly nonlinear problems which arise due to the rapid changes in thermal properties around the freezing point. The use of generic ODE solvers for time stepping is often called the method of lines, although it does not involve any fundamentally new concept compared to conventional approaches. Many pre-programmed ODE solvers are available in the public domain, such as Netlib (2013), NIST (2013b), CASC (2013) or as part of commercial computational software such as Matlab. The advantages of using pre-programmed ODE solvers are:

- they have been extensively tested and optimised
- they may include time step adjustment features which are useful for highly non-linear problems
- the programmer can concentrate on space discretization, i.e. on constructing the matrices in Eq. 5.1.

ODE solvers can be classified into explicit methods (e.g. Euler) or implicit methods. If the capacitance matrix is diagonal, an explicit method will allow the nodal temperatures to be calculated one by one, as has been seen with the Euler method (Eq. 5.38). In other cases, the system of algebraic equations resulting from the ODEs must be solved by a variety of methods, either direct (such as Gauss elimination) or iterative. Scheerlinck et al. (2001) compared the performance of Runge–Kutta methods, implicit backward difference methods and the Crank–Nicolson method, in conjunction with the enthalpy formulation and Kirchhoff transformation (see next sections), and concluded that the order 5 backward difference method and Crank–Nicolson are preferable to explicit methods when accuracy, stability and speed are considered together.

5.3.4 Time Stepping in Structured Grid FDM and FVM

In one dimension, with FDM and structured grid FVM, the $\mathbf{C}$ matrix is diagonal and the $\mathbf{K}$ matrix is tridiagonal (Eq. 5.13), which allow the matrix equation to be solved by the tridiagonal matrix algorithm (TDMA) (Press et al. 1986), a very fast and efficient direct elimination method.

In regular shapes with two or three dimensions (finite cylinders, rods or brick shapes), FDM and FVM can be implemented by applying an orthogonal grid. The nodal temperatures in that case are conventionally written as a 2- or 3-D matrix, $[T_{i,j}]$ or $[T_{i,j,k}]$, each index referring to a dimension (topological notation). To obtain the form in Eq. 5.1 the columns of the temperature matrix must be stacked on each other to form a vector. The resulting $\mathbf{K}$ matrix is no longer tridiagonal since the

Fig. 5.6 Alternating direction algorithm

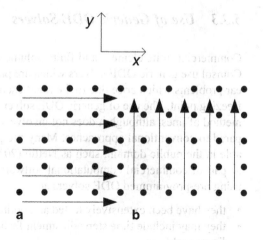

heating rate of each node is influenced by all the neighbouring nodes (four nodes
in two dimensions and six in three dimensions), which are represented by non-
neighbouring members in the **T** vector.

A more efficient approach is the alternating direction method (Peaceman and
Rachford 1955; Douglas 1962) which takes advantage of the regular mesh structure.
The topological matrix notation for nodal temperatures ($[T_{i,j}]$ or $[T_{i,j,k}]$) is retained.
Sweeps are made in each of the x, y and z directions independently in a series of
pseudo-1-D solutions. An example in 2-D using the Crank–Nicolson scheme illustrates the procedure (Fig. 5.6). In the first (x) sweep, we consider one row of nodes
at a time and calculate a set of intermediate temperatures **T***, using a time step of
$\Delta t/2$. The (unknown) intermediate values **T*** are used to express temperature gradients in the x directions and the (known) present values **T** to express temperature
gradients in the y direction:

$$T_{ij}^* = T_{ij} + \frac{\Delta t}{2}(\partial_x T^* + \partial_y T) \qquad (5.45)$$

where $\partial_x T$ is shorthand for $\dfrac{k_+(T_{i+1,j} - T_{i,j}) - k_-(T_{i,j} - T_{i-1,j})}{\rho c \Delta x^2}$ (the heating rate due to

heat flow along the x direction) and similarly for $\partial_y T$. The heat fluxes $\partial_x T^*$ in the x
direction are computed using the intermediate temperature field (i.e. backward differences) while those in the y direction are computed using the existing temperature
field (i.e. explicit). This yields, for each row, a tridiagonal matrix equation that
can be solved for the intermediate temperature vector **T***. In the second (y) sweep,
we consider one column of nodes at a time and write down the discretized Fourier
equation for the next $\Delta t/2$, using the (unknown) new values to express temperature
gradients in the y directions and the (known) intermediate values **T*** to express temperature gradients in the y direction:

$$T_{ij}^{new} = T_{ij}^* + \frac{\Delta t}{2}(\partial_x T^* + \partial_y T^{New}) \qquad (5.46)$$

which again yields a tridiagonal matrix equation for each column that can be solved for $\mathbf{T}^{New}$.

The alternating direction method can also be applied to Lee's three-level scheme for FDM (Cleland and Earle 1979a).

5.4 Dealing with Changes in Physical Properties

The biggest problem in the numerical computation of freezing is how to deal with the sudden changes in material properties around the freezing point: the evolution of latent heat and, to a lesser extent, changes in thermal conductivity.

5.4.1 Latent Heat of Freezing

5.4.1.1 Classification of Methods

In the "classical" freezing problem, also known as the Stefan problem, the freezing process is governed entirely by heat transfer. There is no retardation of freezing due to diffusion or nucleation phenomena. The major problem in the numerical solution of this problem is in dealing with the large latent heat which evolves over a very small temperature range. Except where indicated, the techniques described here apply equally to FDM, FEM and FVM.

Methods to deal with phase change can be divided into *fixed-grid* methods and *moving-grid* methods (Voller 1996). In the latter, the object is divided into a frozen zone and an unfrozen zone. Some nodes, element boundaries or control volume boundaries are put on the freezing front itself and allowed to move with it. The front is tracked by calculating its position at every time interval, using the heat balance equation:

$$-[k\nabla T]_u \cdot \mathbf{n} + [k\nabla T]_f \cdot \mathbf{n} = -\rho L_f \mathbf{v}_f \cdot \mathbf{n} \qquad (5.47)$$

where

$\mathbf{v}_f$	is the velocity of the freezing front,
$\mathbf{n}$	the outward normal unit vector,
L_f	the latent heat of freezing per unit mass of food,
and the subscripts u and f	refer to the heat fluxes on the frozen and unfrozen sides of the front.

The left-hand side of the above equation expresses the net conduction heat gain at the front (the first term is the heat flux entering the front from the unfrozen phase,

Fig. 5.7 Finite volume calculation of temperature histories in a slab of water-ice, showing unphysical stepwise temperature history near the surface. Ten nodes were used

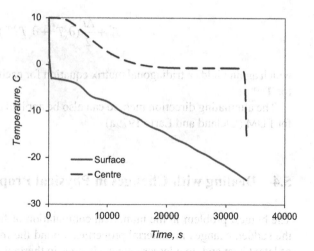

the second is that leaving the front via the frozen phase), while the right hand side expresses the rate of latent heat release.

Moving grid methods can give precise, non-oscillating solutions for the temperature and ice front position. However, they are less flexible than fixed-grid methods because most foods do not have a sharp phase change temperature but freeze gradually, hence it is not clear how the freezing front should be defined. With foods of complex shape, the calculation of the front's position and subsequent grid adjustment can be a complicated affair. Therefore, this chapter will concentrate on fixed grid methods. If desired, the freezing front can still be located in these methods by carrying out an interpolation to locate the position where the freezing temperature T_f applies.

Fixed-grid methods may lead to artificially stepwise temperature histories around the freezing point and stepwise ice front movement (Voller 1996). An example is shown in Fig. 5.7. However, this is much less apparent in foods with their gradual phase change than in a pure substance, and the stepwise characteristic can, in any case, be minimised by reducing mesh size. For 1-D problems, by reporting or plotting the temperature profiles only when a nodal liquid fraction becomes 0.5 (or when a nodal enthalpy is halfway between fully frozen and unfrozen enthalpies), smooth an accurate temperature history and freezing front movement can be obtained (Price and Slack 1954; Voller and Cross 1985). In the following we discuss some of the fixed-grid methods that have been proposed for the freezing of water and foodstuffs.

5.4.1.2 Fictitious Heat Source Methods

Of the fixed-grid methods, some treat latent heat as a source term S_q in Eq. 3.5, separate from the specific heat (Rolph and Bathe 1982; Roose and Storrer 1984; Voller 1990; Voller and Swaminathan 1991). This approach has been used mainly for water and other pure substances, which evolve latent heat at a fixed phase change temperature. With Euler time-stepping, the method is practically equivalent to the explicit enthalpy method to be discussed later, so it is usually applied together with an implicit time stepping scheme. For a simple 1-D control volume (Fig. 5.3), the fully implicit formulation, Eq. 5.42, becomes:

$$-\frac{k_{-}\Delta t}{\Delta x}T_{i-1}^{New} + \left(\rho c \Delta x + \frac{k_{+}}{\Delta x} + \frac{k_{-}}{\Delta x}\right)T_{i}^{New} - \frac{k_{+}\Delta t}{\Delta x}T_{i+1}^{New} = \rho c \Delta x T_{i} + \rho \Delta x L_{f}\left(g_{i}^{New} - g_{i}\right)$$

(5.48)

where

g_i is the frozen fraction of control volume i.

From here on, we revert to the distinction between c which does not include the latent heat and c_{app} which does. Since there is an unknown term g_{i}^{New} on the right-hand side, the equation is no longer linear and has to be solved by an iterative method (Voller 1996).

The source approach is not suitable, or at least not optimal, for foods as their latent heat is evolved over a range of temperature and is, thus, hard to distinguish from sensible heat. Even for pure substances, the phase change can be smeared out over a temperature range $-\delta T$ to $+\delta T$ around the true freezing point, say between -0.1 and $+0.1\,^{\circ}\text{C}$, and the problem treated by one of the other methods considered below. The relative error in the freezing time due to the smearing will be at most of the order $\delta T_{f}/(T_{f}-T_{a})$.

5.4.1.3   Apparent Specific Heat Methods

In the apparent (or effective) specific heat methods, latent heat is merged with sensible heat to produce a specific heat curve with a large peak around the freezing point (Fig. 2.2). Because of the large variations in specific heat, an iteration must be carried out at every step: the specific heat at each node is estimated (say from the present temperatures) and used to calculate the **C** matrix, Eq. 5.35 is solved for the new nodal temperatures, the mean temperature over the most recent time step is calculated, then the specific heat is re-estimated from the specific heat–temperature relationship, and so on. It is difficult to obtain convergence with this technique, and there is always a chance that the latent heat is underestimated ("jumping the latent heat peak"). This happens when a nodal temperature steps over the peak in the apparent specific heat curve: the mean specific heat between the initial and final temperature is then always less than the peak, and the temperature change will, therefore, be overestimated. For this reason, the apparent specific heat method is not recommended.

Early attempts to overcome the latent heat peak jumping problem have consisted of approximating the effective specific heat from recent nodal temperature changes or from local temperature gradients (Comini and Del Giudice 1976; Morgan et al. 1978; Comini et al. 1989; Lemmon 1979; Cleland et al. 1984). Although these methods avoid the need for iteration, their theoretical background is weak and performance not entirely satisfactory (Pham 1995).

A rigorous solution requires an iterative method such as Gauss–Seidel or Newton to solve the system of nonlinear equations generated by the PDEs. Under- or over-relaxation may be employed to accelerate convergence. This approach is used by generic finite element software such as Comsol, Ansys or Abaqus, and is very time consuming, especially for 3-D problems where a large number of nodes are involved.

5.4.1.4   Enthalpy Methods

In the enthalpy method, the heat conduction equation, Eq. 3.5, is written in the form

$$\rho\frac{\partial H}{\partial t} = \nabla(k\nabla T) \tag{5.49}$$

where the source term has been dropped and

H is the (specific) enthalpy:

$$H = \int_{T_{ref}}^{T} c_{app}\, d\theta \tag{5.50}$$

T_{ref} is an arbitrarily chosen reference temperature and c_{app} is the apparent specific heat. After the usual FDM, FEM or FVM manipulations, we obtain the matrix equations

$$\mathbf{M}\frac{d\mathbf{H}}{dt} + \mathbf{KT} = \mathbf{f} \tag{5.51}$$

where

M   is a mass matrix which remains constant with time, and
H   is the vector of nodal enthalpies.

To apply the enthalpy method, the functional relationship $T(H)$ must be available and programmed into the computer. In FDM, FVM and lumped mass FEM, the solution of Eq. 5.51 by Euler's method (or any other explicit time stepping method) is very simple, since **M** is a diagonal matrix, and each new nodal enthalpy will be given by:

$$H_i^{New} = H_i + \frac{\delta t}{M_{ii}}\left[\sum_{j=1}^{N}(K_{ij}T_iT_j) + f_i\right],\ i = 1\text{ to }N \tag{5.52}$$

where all the terms on the right are known (present) values. This method was first proposed for FDM by Eyres et al. (1946). The new nodal enthalpies are calculated one by one from the present temperature field, then the new nodal temperatures are calculated from the H–T relationship and substituted back into Eq. 5.52, and so on.

To obtain an exact solution to any implicit ($\alpha > 0$) solution of Eq. 5.51, iteration must be carried out at every time step. The enthalpy change vector $\Delta \mathbf{H} \equiv \mathbf{H}^{New} - \mathbf{H}$ over the present time step is iteratively adjusted until the residual vector $\mathbf{r} = \mathbf{M}\Delta\mathbf{H}/\Delta t + \overline{\mathbf{K}}\overline{\mathbf{T}} - \overline{\mathbf{f}}$ becomes zero to within an acceptable tolerance. A successive substitution scheme such as Gauss–Seidel can be used:

$$\mathbf{H}^{New} = \mathbf{H} + \Delta t \cdot \mathbf{M}^{-1}(\overline{\mathbf{K}}\overline{\mathbf{T}} - \overline{\mathbf{f}}) \tag{5.53}$$

$$\mathbf{T}^{New} = \mathbf{T}(\mathbf{H}^{New}) \tag{5.54}$$

Convergence with this type of scheme tends to be very slow and various overrelaxation schemes have been proposed (Voller et al. 1990). A better approach is to use a Newton–Raphson iteration, where the following equation is solved iteratively for the enthalpy change vector $\Delta\mathbf{H}$ (m being the iteration counter) until the residual $\mathbf{r}$ is reduced to an acceptably small value:

$$\mathbf{J}\Delta\mathbf{H}^{m+1} = \mathbf{J}\Delta\mathbf{H}^m - \mathbf{r}^m \tag{5.55}$$

$$\mathbf{J} = \frac{d\mathbf{r}}{d\mathbf{H}^{New}} = \frac{\mathbf{M}}{\Delta t} + \mathbf{K}\frac{\partial \overline{\mathbf{T}}}{\partial \mathbf{H}^{New}} - \frac{\partial \overline{\mathbf{f}}}{\partial \mathbf{H}^{New}} \tag{5.56}$$

For example, if the Crank–Nicolson scheme is used ($\overline{\mathbf{T}} = (\mathbf{T} + \mathbf{T}^{New}/2)$, $\overline{\mathbf{f}} = (\mathbf{f} + \mathbf{f}^{New})/2)$, the Jacobian $\mathbf{J}$ becomes:

$$\mathbf{J} = \frac{\mathbf{M}}{\Delta t} + \frac{1}{2}\left(\mathbf{K}\frac{\partial \mathbf{T}^{New}}{\partial \mathbf{H}^{New}} - \frac{\partial \mathbf{f}^{New}}{\partial \mathbf{H}^{New}} \right) \tag{5.57}$$

Other iterative solution methods are available (see for example Press et al. 1986).

The discretization of highly non-linear problems such as the phase change problem with Galerkin FEM poses serious difficulties from the physical point of view. The Galerkin approach assumes that temperature is distributed over the element according to the shape function, i.e. $T = \mathbf{N}^{T}\mathbf{T}$ where $\mathbf{T}$ is the vector of nodal temperatures. Since H is a non-linear function of T, it cannot be assumed that H is distributed according to $H = \mathbf{N}^{T}\mathbf{H}$ as well, where $\mathbf{H}$ is the vector of nodal enthalpies. In fact, this interpolation will be very inaccurate around the freezing point (see Fig. 5.8 where the nodal temperatures in a 1-D element are just above and below the freezing point). Yet, in the enthalpy method, this assumption has to be made when transforming Eq. 5.49 into a Galerkin FEM equation. In the effective specific heat method, the Galerkin FEM user is faced with how to calculate an effective specific heat over both time (Δt) and space (the element's domain): numerical averaging methods, which use some sampling procedure over the element's domain, are inaccurate when $c_{app}(T)$ has a very sharp peak which may lie entirely outside the

Fig. 5.8 Temperature and
enthalpy profiles in a finite
element

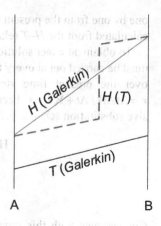

A   B

sampling points. For these reasons, the use of lumped mass FEM (diagonal mass
matrix) is highly recommended over Galerkin FEM, especially when the freezing
point is sharp.

5.4.1.5 Pham's Temperature Correction Method (Quasi-enthalpy Method)

Pham (1985b) proposed a simple correction to the specific heat formulation which,
like enthalpy methods, is effective in dealing with the latent heat peak, but avoids
the need for iteration. The method was first applied to FDM, but was subsequently
extended to lumped mass FEM (Pham 1986c) and to Galerkin FEM (Comini et al.
1989). The method essentially consists of adding a specific heat estimation step and
a temperature correction step.

a. Specific heat estimation step: For each time step, the nodal enthalpy changes
   are first estimated from the incoming heat fluxes, using the present temperature field $\mathbf{T}$:

$$\Delta \mathbf{H} = \mathbf{M}^{-1}\left(\mathbf{KT} - \mathbf{f}\right)\Delta t \qquad (5.58)$$

and an effective apparent specific heat over the time interval can be estimated from:

$$c_{app,i} = \frac{\Delta H_i}{f_T(H_i + \Delta H_i) - T_i} \qquad (5.59)$$

where

$f_T(H_i + \Delta H_i)$ is the temperature corresponding to enthalpy value $H_i + \Delta H_i$.

These effective specific heats are then substituted into the matrix $\mathbf{C}$ in Eq. 5.35,
which is then solved once only per time step, to produce nodal temperatures T_i^{New}.

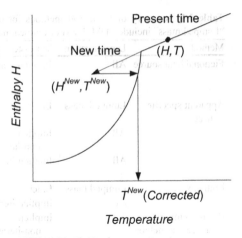

Fig. 5.9 Illustration of Pham's temperature correction method

b. Temperature correction step: To avoid "latent heat peak jumping", the new nodal temperatures are then corrected according to:

$$H_i^{New} = H_i + c_{app,i}\left(T_i^{New} - T_i\right) \tag{5.60}$$

where

H_i is the present enthalpy of node i.

The new nodal temperatures are then corrected to:

$$T_i^{New}(corrected) = f_T(H_i^{New}) \tag{5.61}$$

Further investigations (Pham 1995) indicated that the temperature correction step is the crucial step in this method, while the preliminary specific heat estimation step is of lesser importance and the specific heat value at the present nodal temperature T_i can also be used. The physical basis of the temperature correction is illustrated in Fig. 5.9. It can be seen from the last equation that this is basically an enthalpy method, since it is really the nodal enthalpy changes that are calculated at each time step. As in other enthalpy methods, the functional relationship $f_T(H)$ is needed.

Using well-known test problems, Pham (1995) compared ten of the most advanced fixed-grid finite element methods at the time (after eliminating several others) in terms of accuracy, time interval for convergence, heat balance error (percentage difference between heat flows through boundaries and total heat gain of product—a measure of whether the latent heat load peak has been missed) (Sect. 5.5), and computing time as measured by the number of matrix inversions required. The test problems use both materials with a sharp phase change (heat released over 0.01 K) and a material with food-like properties. He concluded that the (non-iterative) lumped mass FEM with Pham's quasi-enthalpy method performed best in

Table 5.1 Summary of fixed grid methods for dealing with latent heat in numerical models. "Lumped mass" include FDM, FVM or lumped mass FEM

Method	Discretization	Time-stepping	Material	Accuracy	Speed
Fictional heat source	All	Implicit iterative	Sharp or well-defined phase change	Good	Slow
Apparent specific heat	Lumped mass	Euler	Gradual phase change	Poor	Slow
	All	Implicit non-iterative	Gradual phase change	Poor	Slow
	All	Implicit iterative	Gradual phase change	Poor	Slow
Enthalpy	Lumped mass	Euler	All materials	Good	Slow
	All	Implicit iterative	All materials	Good	Slow
Pham's temperature correction method	All	Implicit non-iterative	All materials	Good	Fast

terms of most of the above criteria. Voller (1996) also concluded that this method is an "excellent scheme" for fixed grids.

Because no iteration is involved, strict energy conservation is not maintained by the above methods (heat fluxes are calculated using the uncorrected temperatures while nodal enthalpy changes are based on the corrected temperature). This apparent defect can be turned to advantage in estimating the accuracy of the results: the overall heat balance (relative difference between the boundary flux, integrated over all boundaries and entire freezing time, and the total change in the heat content of the food) can serve as a useful indication of whether the time step is sufficiently small: a heat balance error of less than 1 % (preferably less than 0.1 %) generally indicates that convergence has been reached. Iterative enthalpy methods ensure a good heat balance at all time steps, but this does not guarantee an accurate solution because at large time intervals the non-linearities may be "smoothed over".

Table 5.1 summarises the main characteristics and applicability of the fixed grid methods discussed in this chapter.

5.4.2 Variable Thermal Conductivity

The rapid change in thermal conductivity around the freezing point contributes to the difficulty in the numerical modelling of phase change. In computing the heat flux between nodes $i+1$ and i, it is unclear what value should be used for k: the thermal conductivity calculated at the mean temperature $(T_{i+1}+T_i)/2$, the average thermal conductivities $(k_{i+1}+k_i)/2$, or some other combination such as the series model, $(0.5/k_{i+1}+0.5/k_i)^{-1}$. A more rigorous formulation is obtained by using the Kirchhoff transformation (Comini et al. 1989; Voller and Swaminathan 1991):

$$w = \int_{T_{ref}}^{T} k d\theta \qquad (5.62)$$

$$dw = kdT \qquad (5.63)$$

which when substituted into the heat conduction equation (Eq. 3.5) gives

$$\frac{\rho c}{k} \frac{\partial w}{\partial t} = \nabla^2 w + S_q \qquad (5.64)$$

Note that c should be replaced by c_{app} if the apparent specific heat method is used. The ratio $\rho c/k$ is a property of the material which depends on temperature, and therefore on w only. This technique groups all the nonlinearities into a single factor $\rho c/k$, after which the equation can be solved by FDM, FEM or FVM using an apparent specific heat method (with T replaced by w and c replaced by $\rho c/k$). Voller (1996) found that the Kirchhoff transformation gave a much more accurate prediction of freezing front position than the use of an average k value.

Alternatively, the left hand side is written in terms of enthalpy (Comini and del Giudice 1976; Fikiin 1996; Scheerlink et al. 2001):

$$\rho \frac{\partial H}{\partial t} = \nabla^2 w + S_q \qquad (5.65)$$

which can be then solved by any enthalpy type method or Pham's quasi-enthalpy method, with T replaced by w. To solve the equation, a functional relationship must first be derived between H and w. Scheerlink (2000, 2001) found that the Kirchhoff transformation leads to a significant reduction in computation time when using an iterative method, because the $\mathbf{C}$ and $\mathbf{K}$ matrices becomes constant and do not have to be recomputed.

With composite materials, the Kirchhoff transformation requires additional equations to describe the boundaries between different materials. For example, when two adjacent FEM elements made of different materials share the same nodes, the values of w at these nodes will be different depending on whether they are viewed from one element or the other. Each node that is shared by two materials A and B must be treated as two separate nodes, and a (generally nonlinear) equilibrium relationship must be added to the equation system for each of these node pair:

$$T\left(w_i^A\right) = T\left(w_i^B\right) \qquad (5.66)$$

The use of the Kirchhoff transformation is problematic when mass transfer causes significant changes in moisture content. Since thermal properties then depend on moisture content as well as temperature, there will be no unique relationship between w and H.

A complication which is not often taken into account is anisotropy. Fresh foods often have fibrous or otherwise anisotropic structures. Even in an isotropic matrix, the dendritic growth of ice may cause thermal conductivity to be direction-dependent. Anisotropy can be easily taken into account in numerical models (simply replace $k\nabla T$ by $\mathbf{k}.\nabla T$ in Eq. 3.5, where $\mathbf{k}$ is the thermal conductivity vector). However, due to lack of accurate property data it is rarely considered.

5.4.3 Variable Density due to Thermal Expansion

Water expands by about 9 % upon freezing, causing a change in the density of foodstuffs. This means that the heat conduction equation (Eq. 3.5) no longer applies, since there are material movements (convection) that must be taken into account. To model the expansion accurately, we have two options: either a strictly Eulerian mesh which is fixed in space, or a Lagrangian (embedded) mesh which expands with the product. With the former, a convection term arises due to the movement of expanding matter across the (fixed) boundaries of the control volume or element. The Eulerian approach can be used in the case of a freezing liquid where convection is present anyway. For solid foods, a Lagrangian mesh is much more convenient as no convection term will be involved.

In a Lagrangian or embedded mesh, each control volume or element encloses a fixed *mass* of product (excluding any diffusional effect). Consider again the finite volume grid in Fig. 5.3. Let x denote the *original* space coordinate and ξ the expanded coordinate. The heat flux from node $i-1$ to node i is:

$$k_-\frac{T_{i-1}-T_i}{\Delta\xi} = k_-\frac{T_{i-1}-T_i}{(1+e_x)\Delta x} \qquad (5.67)$$

where

e_x is the fractional linear expansion in the x direction.

At the same time, the cross-section area for heat flow into the control volume has expanded by $(1+e_y)(1+e_z)$. The heat *flow* into the control volume (which will have expanded in all directions, even in a 1-D model) must, therefore, be multiplied by $(1+e_y)(1+e_z)/(1+e_x)$, or (assuming isotropic expansion) $(1+e_v)^{1/3}$, where e_v is the volumetric expansion. The mass of the control volume remains at $\rho_o\Delta x$ per unit original area, where ρ_o is the original (unfrozen) density. The heat balance over a control volume, Eq. 5.14, then becomes:

$$\rho_o c\Delta x\frac{\partial T_i}{\partial t} = (1+e_v)^{1/3}\left[k_-\frac{T_{i-1}-T_i}{\Delta x}+k_+\frac{T_{i+1}-T_i}{\Delta x}\right] \qquad (5.68)$$

In other words, the thermal conductivity k must be replaced by an expansion-corrected conductivity $(1+e_v)^{1/3}k$. The above equation can also be obtained by discretizing the following equation, which is obtained by putting $\rho = \rho_o/(1+e_v)$ and $\nabla = \nabla_o(1+e_v)^{-1/3}$ in Eq. 3.5 and discarding the source term:

$$\rho_o c \frac{\partial T}{\partial t} = (1+e_v)^{1/3} \nabla_o \cdot (k \nabla_o T) \tag{5.69}$$

where

$\nabla_o = (\partial/\partial x, \partial/\partial y, \partial/\partial z)$ is the gradient with respect to the original coordinates.

Because $(1+e_v)^{1/3}-1 \approx e_v/3$ (usually 0.02 or less) is often much smaller than the uncertainties in k values, and because isotropic expansion cannot always be assumed, most works in the field have ignored the effect of freezing expansion on heat transfer. As can be seen from Eq. 5.68, this is acceptable provided that the initial (unfrozen) density ρ_u is used throughout the calculations (for freezing). If the frozen density is used on the left a significant error may occur.

5.5 Convergence and Accuracy of Numerical Methods

Numerical methods are generally considered the most accurate class of methods for solving physical problems, when analytical solutions are not available. However, it is important to realise their limitations. For a numerical solution to be completely reliable, the following conditions must be fulfilled:

- The physics of the problem must be thoroughly understood and all non-negligible factors taken into account.
- The effects of all the influencing factors are described by well-founded mathematical relationships. For example, the heat conduction equation (Eq. 3.5) is one such relationship, but present turbulent flow models, which rely on highly empirical equations, are not, so any model that involves turbulent flow must be experimentally verified.
- The mathematical relationships must be translated into a numerical model by suitable methods. For example, several numerical methods have been proposed to treat the latent heat peak (Sect. 5.4.1) but not all are accurate or consistent.
- The model's inputs—in this case the thermal properties of the food and the environmental conditions—must of course be accurate.
- The numerical program must have been validated against analytical solutions if possible.
- The calculations must have converged.

The question of convergence is an important consideration and will be considered in more detail. The accuracy of a numerical solution depends both on the quality and density (fineness) of the space mesh (see Sect. 5.2.4). Commercial meshing software generally automatically ensure reasonable mesh quality and may offer automatic mesh adaptation to take care of steep gradients, but the fineness of the mesh is often still controlled by the user.

To ensure that the mesh is sufficiently fine, the numerical simulation should be run with two different mesh sizes, one being, say, twice as fine as the other, and the results compared: if they are similar or identical then the mesh is satisfactory. For freezing calculations a minimum of 10 nodes should be used for the half slab.

Similarly, convergence tests should be carried out to ensure that the time step is sufficiently short. If a constant time step is used, the program should be run with decreasing time step size until the results converge to constant values. If adaptive time-stepping is used (as is the case in many commercial software), where the length of each time step is adjusted until the results converge to within a certain tolerance, then different values for the tolerance limit should be tried.

A simple test for checking that the time step is sufficiently small is the heat balance check. The heat flow across the product surface is calculated at each time step and accumulated over the whole run. At the end of the run this accumulated heat flow is compared with the total change in enthalpy of the product, obtained by integration. If the discrepancy is, say, 0.1 % or less then the time step can be considered sufficiently small. This method may fail if there are large fluctuations in the environmental temperature—for example, it obviously will not work if the product is cooled then heated back to the original temperature—and also does not say anything about whether the space mesh is satisfactory.

Fine meshes and small time steps will increase computation time. This can be alleviated somewhat by taking advantage of any symmetry that exists. If boundary conditions are uniform or symmetrical, only half a slab, a quarter of a rectangular rod or an eighth of a brick should be modelled. Using axial symmetry, a finite cylinder can be treated as a 2-D object (a rotated rectangle).

5.6 Summary and Recommendations

- If the user is writing his/her own numerical program for simulating the freezing or thawing of foods, the recommended approach is FVM or lumped mass FEM. Conventional distributed mass FEM may miss the latent heat peak if the latter is very sharp and the mesh is coarse, while FDM is less efficient at handling boundary conditions.
- To deal with the latent heat load, either Pham's temperature correction method (Sect. 5.4.1.5) or the explicit enthalpy method (Sect. 5.4.1.4) can be used. The former is faster and should be used if the program is to be run many times, while the latter is simpler to program.
- To deal with variable thermal conductivity, the Kirchhoff transformation is recommended (Sect. 5.4.2). However it may be difficult to apply when moisture movement strongly influences heat transfer.

- To deal with variable density, a Lagrangian or embedded mesh is recommended (Sect. 5.4.3). However, most past works have ignored density variation, which cannot be accounted for accurately unless the relative expansion in each direction is known.
- For multidimensional geometries the alternating direction method is recommended, provided a structured grid is used (Sect. 5.3.4)
- Generic commercial FVM and FEM software packages take care of most of the calculation details but can be very slow for 3-D problems.
- With commercial numerical software, the user has little choice in the methodology. An iterative apparent specific heat method is almost always used and the user's main concern is how to enter the parameters, in particular the specific heat-temperature function, accurately. The best way is to enter the enthalpy-temperature relationship as a tabular or other function, and define the specific heat as its temperature derivative. The user's choices are usually restricted to choosing the mesh size and setting the tolerance limits for the iterative solution. If an option exists for lumped capacitance (diagonal mass matrix) it should be chosen.
- Numerical simulation is able to simulate the freezing and thawing of solid foods with great accuracy as long as heat conduction is the only or dominant mechanism. However, certain precautions are necessary to ensure the accuracy of the results

⚠ CAUTION

- In some publications, a "control volume FDM" is used. This is, in fact, the structured-grid version of FVM described in this work. In some implementations, the surface control volume may be only half the size of the other control volumes and the corresponding node is placed at the surface, i.e. to one side of its control volume instead of being at its centre. As long as the mesh is sufficiently fine this does not cause serious problems.

Chapter 6
Modelling Coupled Phenomena

6.1 Introduction

So far we have assumed that heat conduction is the only mechanism governing the freezing of foods, according to Eq. 3.5. In practice, however, phase change almost never happens in isolation and other phenomena take place in parallel with heat transfer. Moisture movement happens on the macro scale and at crystalline and cellular level. Supercooling is the rule rather than the exception. The use of high pressure will influence nucleation and phase change. Complex stresses and strains develop that may cause cracking, texture change, and other alterations that affect food quality.

6.2 Combined Heat and Mass Transfer

In food freezing, heat transfer is always accompanied by mass transfer and the latter may have important implications for weight loss and product quality. We will concentrate on the transfer of moisture only, which is the most common situation, although solute transfer also happens in immersion freezing.

When mass transfer occurs, conduction is not the only mode of heat transfer. Thermal energy is also conveyed by the diffusing substance, necessitating the addition of a second transport term. This can most easily be expressed with the enthalpy form of the heat transport equation:

$$\frac{\partial (\rho H)}{\partial t} = \nabla(k\nabla T) + \nabla(H_w \dot{m}) \tag{6.1}$$

where $\dot{m}$ is the mass flux and H_w the enthalpy of the diffusing substance, and density changes due to mass flux imbalance is taken into account. The mass flux is assumed to follow Fick's law:

$$\dot{m} = -\rho_{ds} D_w \nabla W \tag{6.2}$$

Q. T. Pham, *Food Freezing and Thawing Calculations*,
SpringerBriefs in Food, Health, and Nutrition, DOI 10.1007/978-1-4939-0557-7_6,
© The Author 2014

where ρ_{ds} is the mass concentration of dry solid, W is the *dry basis* mass fraction of the diffusing substance (kg/kg dry solid) and D_w its (effective) diffusivity. The governing equation for mass transfer is therefore:

$$\frac{\partial W}{\partial t} = \nabla(D_w \nabla W) \tag{6.3}$$

Mechanical effects (gravity, pressure gradient) have been ignored, as well as the mass diffusion due to temperature gradient (Soret effect) and heat diffusion due to concentration gradient (Dufour effect). The second term in the heat transport equation can usually be neglected in dense foods due to the very slow moisture diffusion rate, but not in porous foods, where evaporation or condensation may occur causing large changes in H_w.

In the freezing of meat and other dense (non-porous) foods, water evaporates from the surface and is replenished by deep water diffusing towards the surface, until freezing occurs. Thereafter the water sublimes from the ice front at or near the surface, but there is no significant water movement in the food. Mass transfer occurs in a rather thin layer near the surface only, in contrast to heat transfer which happens throughout, and the main problem is how to deal with the different scales effectively. In the freezing of porous food, such as bread and dough, moisture movement continues right through the freezing process deep inside the food. Here, the heat and mass transfer scales are similar. Yet another type of mass transfer occurs in whole biological tissue between the intra- and extracellular spaces. Each situation presents a different set of challenges that may require a special modelling approach.

6.2.1 Mass Transfer During the Freezing of Dense Foods

Moisture in foods can simultaneously exist in several phases: vapour, "free" liquid, ice, and various types of "bound" moisture, each of which has its own diffusion rate. Because the phases are in intimate contact, the different phases are usually assumed to be in thermodynamic equilibrium with each other (this may not be true if supercooling and kinetic effects are present). However, due to lack of data, it is commonly assumed that moisture movement in foods can be described by a single-phase diffusion equation (Eq. 6.3), with an effective diffusivity D_w. This equation is of the same form as the heat conduction equation (Eq. 3.5), and can be solved by the same methods (FDM, FEM, or FVM). In dense food, moisture diffusion is very slow and its contribution to heat transport can be neglected, hence the second term in Eq. 6.3 can be neglected, except at the evaporating surface itself.

The problems to be considered are the changes in boundary conditions and the differences in scale between heat and mass transfer. During the precooling phase (prior to surface freezing), water evaporates from the surface and is replenished by moisture diffusing from the inside to the surface. This balance is expressed by:

$$\rho_{ds} D_w \left(\frac{\partial W}{\partial n} \right)_s = -k_g (P_s - P_a)$$

$$\tag{6.4}$$

where ρ_{ds} is the density of the dry solid component, n the unit normal vector, k_g the mass transfer coefficient (kg·m^{-2} s^{-1} Pa^{-1}), P_s the water partial pressure at the food surface, and P_a that in the surroundings. The left hand side is the moisture flux towards the surface as expressed by Fick's law, while the right hand side is the water vapour flux from the surface to the surrounding air. P_s is related to the surface moisture by:

$$P_s = \alpha_w P^*(T_s)$$ (6.5)

where $P^*(T_s)$ is the saturated water vapour pressure at the surface temperature T_s and α_w the surface water activity. The latent heat of vaporization λ_v must be taken into account in the boundary condition of the heat conduction equation:

$$-k\left(\frac{\partial T}{\partial n}\right)_s = h(T_s - T_a) + \lambda_v k_g (P_s - P_a)$$ (6.6)

where h is the surface heat transfer coefficient, which may include radiation effects. The boundary conditions for both heat and mass transfer become non-linear since they contain a term P_s, which is a function of temperature and moisture. If Euler time stepping is used, this poses no particular problem, since all variables are calculated explicitly from known conditions. If a non-iterative implicit stepping method is used, P_s can be linearised around the present temperature and moisture value (Davey and Pham 1997):

$$P_s = a_1 + b_1 T_s$$ (6.7)

$$P_s = a_2 + b_2 W_s$$ (6.8)

If an iterative method is used, the nonlinear boundary conditions are solved iteratively with the other equations.

A further complication is caused by the differences in scale between heat and mass transfer. Moisture diffusivity in dense foods is typically of the order $D_w \approx 10^{-10}$ m^2 s^{-1}, while thermal diffusivity is of order $k/\rho c \approx 10^{-6}$ m^2 s^{-1}, which means that by the time the freezing process is completed, only a very thin layer near the surface has undergone any noticeable change in moisture content. To model moisture movement accurately, therefore, would require an extremely fine grid, which (because of the high thermal diffusivity), would in turn require extremely small time intervals to avoid a large value of $\rho c \Delta t / \Delta x^2$, which would cause excessive oscillation and inaccuracies in the temperature field. Pham and Karuri (1999) overcame this difficulty by using a two-grid method, where a second, very fine, one dimensional finite volume surface grid was used to model the mass transfer. At each time step, the heat flow equation was solved first using the first grid, then the mass transfer equation using the second grid. The approach was implemented by Trujillo (2004) in modelling the chilling of a beef side, using the FVM-based CFD software FLUENT. In this case the mass transfer grid was incorporated as a user-defined function.

Once the surface has frozen (at an initial freezing point determined thermodynamically by the surface water activity), water becomes immobilized and internal diffusion stops. Moisture then sublimes, at first from the surface, then through a layer of desiccated food that gradually thickens as the ice front recedes, at a rate determined by (Pham and Willix 1984; Pham and Mawson 1997)

$$\dot{m} = \frac{P_{sat}(T_s) - P_a}{1/k_g + \delta/D_\delta} \tag{6.9}$$

where δ is the desiccated thickness and D_δ the effective diffusivity of water through it. The problem was modelled for 1-D geometry using a front-tracking finite difference method (Campañone et al. 2001). The dehydrated zone was modelled by a flexible grid with distance increments increasing proportionately to the depth of the freezing zone. The un-dehydrated zone (both frozen and unfrozen) was modelled by a fixed grid, except that the last node moved with the sublimating interface (and hence, the last space increment of the un-dehydrated zone decreased with time). An apparent heat capacity method appeared to have been used to deal with the freezing front.

Because the desiccated layer is normally very thin in the freezing of dense foods, modelling it numerically requires a very thin grid. In fact, at the moment when freezing starts, δ may be zero, so an infinitely fine grid is required, which would cause some difficulty. (Unfortunately, Campañone et al. did not mention how this was handled.) In the author's view, when the rate of sublimation is very slow and the desiccated layer very thin (thinner than, say, half the thickness of a control volume), we can assume pseudo-steady state (i.e. the water vapor profile in the desiccated layer is as if the sublimating front was stationary), and it is sufficiently accurate to use an ODE approach:

$$\frac{d\delta}{dt} = \frac{\dot{m}_w}{\rho_{ds} W(\delta)} \tag{6.10}$$

where $W(\delta)$ is the moisture content at depth δ. This equation can be integrated over each time interval and the resulting value of δ substituted into Eq. 6.9 to calculate the surface mass flux, which can then be used in the boundary condition for the heat and mass transfer PDEs. For porous foods, however, the thickness of the desiccated layer may be much thicker and so it would be more appropriate to model it with a FDM, FEM or FVM grid.

6.2.2 Mass Transfer During Air Freezing of Porous Foods

With porous food, mass diffusion is much faster than in dense foods due to vapour diffusion in the pores. Water evaporates from the warm inner parts of the food and diffuses towards the outside. When the freezing point is reached, the vapour

Fig. 6.1 Evaporation-condensation process in a porous food

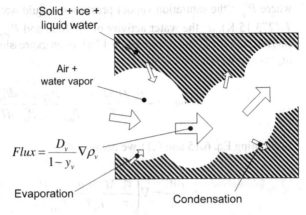

Solid + ice + liquid water

Air + water vapor

$Flux = \dfrac{D_v}{1-y_v}\nabla\rho_v$

Evaporation

Condensation

condenses into ice. This situation has been modelled by van der Sluis (1993) and Hamdami et al. (2004a, b, c) for bread freezing. The presence of three phases is a complicating factor and previous works have made various approximations, so a more rigorous model will be developed here.

The moisture balance can be written:

$$\rho_{ds}\frac{\partial W}{\partial t} = -\nabla\dot{m}_w \qquad (6.11)$$

where ρ_{ds} is the dry solid density, W the total moisture content per unit mass of dry solid and $\dot{m}_w$ the moisture flux. We shall assume that the moisture diffuses mainly in the vapour phase, where the flux based on total cross-sectional area is given by (Fig. 6.1):

$$\dot{m}_w = -\frac{\phi_g}{\tau}\frac{M_w CD_v}{1-y_w}\nabla y_w \qquad (6.12)$$

where C is the molar concentration of water in the gas phase, ϕ_g the void fraction, τ the tortuosity factor of the diffusion path, y_w the mole fraction of water vapour in the pores, and D_v the diffusivity of water vapour in air. In this expression $-M_w CD_v\nabla y_w$ is the unrestricted water vapour diffusion flux from Fick's first law, $1/(1-y_w)$ is a correction factor for bulk movement (Bird et al., 1960) and ϕ_g/τ a geometric factor. The total gas molar concentration is, from the ideal gas law:

$$C = \frac{P_{atm}}{R_g T} \qquad (6.13)$$

and the water vapour mole fraction is:

$$y_w = \frac{a_w P^*_{liq}}{P_{atm}} \qquad (6.14)$$

where P_{liq}^* is the saturation vapour pressure of liquid water (or supercooled water if $T < 273.15$ K), a_w the water activity of the food and P_{atm} the atmospheric pressure. Substituting for C and y_w into Eq. 6.12 gives an expression for the water vapour flux $\dot{m}_w$ (assuming constant a_w):

$$\dot{m}_w = -\frac{\phi_g}{\tau}\frac{D_v M_w a_w}{R_g T}\frac{P_{atm}}{P_{atm} - a_w P_{liq}^*}\frac{dP_{liq}^*}{dT}\nabla T \tag{6.15}$$

Combining Eq. 6.15 and 6.11 we obtain:

$$\rho_{ds}\frac{\partial W}{\partial t} = \nabla\left(\frac{\phi_g}{\tau}\frac{M_w D_v a_w}{R_g T}\frac{P_{atm}}{P_{atm} - a_w P_{liq}^*}\frac{dP_{liq}^*}{dT}\nabla T\right) \tag{6.16}$$

When ice is present, it exists as a pure phase, in contrast to the liquid water which normally will contain solutes that depress a_w, and therefore the vapour pressure of the ice-containing (frozen) food is the same as that of ice. Equation 6.16 then becomes:

$$\rho_{ds}\frac{\partial W}{\partial t} = \nabla\left(\frac{\phi_g}{\tau}\frac{M_w D_v}{R_g T}\frac{P_{atm}}{P_{atm} - P_{ice}^*}\frac{dP_{ice}^*}{dT}\cdot\nabla T\right) \tag{6.17}$$

where P_{ice}^* is the vapour pressure of ice.

The continuity condition at the freezing point gives a_w at temperatures above freezing:

$$a_w = \frac{P_{ice}^*(T_f)}{P_{liq}^*(T_f)} \tag{6.18}$$

Figure 6.2 illustrates the relationship between the water vapour pressures of food, pure liquid water and ice. As temperature falls below T_f, ice forms and the liquid solution becomes more and more concentrated, lowering the activity with respect to supercooled liquid.

The heat transport equation differs depending on whether ice is present. If ice is not present, the flux imbalance represents an evaporation or condensation from/to liquid water, causing a heat source $-\lambda_{vl}\nabla\dot{m}_w$ where $\lambda_{vl} = H_v - H_l$ is the latent heat of evaporation of water. The heat balance then gives:

$$\rho c\frac{\partial T}{\partial t} = \nabla(k\nabla T) - \lambda_{vl}\nabla\dot{m}_w \tag{6.19}$$

where k is the thermal conductivity which may be calculated from one of the models in Chap. 2, and c is the specific heat. ρc depends on moisture and can be calculated from:

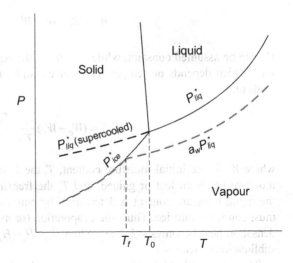

Fig. 6.2 Phase diagram of water. The *dotted red line* gives the water vapour pressure in the food

$$\rho c = \rho_s (c_{ds} + W c_l) \qquad (6.20)$$

where c_{ds} is the specific heat of dry solid and c_l that of liquid moisture. In this equation the specific heat of bound water is assumed to be the same as that of free liquid water, c_l, a commonly made assumption. Substituting for $\dot{m}_w$ from Eq. 6.15 gives:

$$\rho c \frac{\partial T}{\partial t} = \nabla \left[\left(k + \frac{\phi_g}{\tau} \frac{\lambda_{vl} D_v M_w a_w}{R_g T} \frac{P_{atm}}{P_{atm} - a_w P_{liq}^*} \frac{dP_{liq}^*}{dT} \right) \nabla T \right] \qquad (6.21)$$

which can be written as:

$$\rho c \frac{\partial T}{\partial t} = \nabla \left[\left(k + k_{eva-con} \right) \nabla T \right] \qquad (6.22)$$

where $k_{eva-con}$ is the effective thermal conductivity due to evaporation-condensation (De Vries 1952, 1958, 1987):

$$k_{eva-con} = \frac{\phi_g}{\tau} \frac{\lambda_{vl} D_v M_w a_w}{R_g T} \frac{P_{atm}}{P_{atm} - a_w P_{liq}^*} \frac{dP_{liq}^*}{dT} \qquad (6.23)$$

When ice is present, the situation is more complicated since there are three phases in equilibrium. Due to local changes in temperature and moisture, the amounts of water vapour, liquid water and ice may all change. We will neglect the contribution of water vapour to the total moisture mass due to its low density and assume that moisture consists of bound moisture W_b, free liquid moisture W_l and ice W_{ice}:

$$W = W_b + W_l + W_{ice} \tag{6.24}$$

W_b can be assumed constant, while $W_l = W_{l,eq}$, the equilibrium unfrozen free moisture, which depends on temperature. For example, for an ideal solution Eq. 2.16 leads to:

$$W_{l,eq} = (W_0 - W_b)\frac{T_f - T_0}{T - T_0} \tag{6.25}$$

where W_0 is the initial moisture content, T_f the *initial* freezing point, before any moisture has been lost or gained, and T_0 the freezing point of pure water. Since the liquid moisture content is determined by equilibrium, any condensing vapour must condense into ice. Thus, the evaporation (or more exactly sublimation)-condensation heat becomes $-\lambda_{vi}\nabla \dot{m}_w$ where $\lambda_{vi} = H_v - H_{ice} = \lambda_{vl} + \lambda_f$ is the latent heat of sublimation of ice.

There is a further latent heat source caused by liquid water freezing or ice melting to maintain the liquid water content at equilibrium with ice:

$$-\lambda_f \rho_{ds}\frac{dW_{l,eq}}{dt} = -\lambda_f \rho_{ds}\frac{dW_{l,eq}}{dT}\frac{dT}{dt} \tag{6.26}$$

where $dW_{l,eq}/dT$ can be obtained from an equilibrium relationship such as Eq. 6.25. The heat equation 6.21 now becomes:

$$\rho c\frac{\partial T}{\partial t} = \nabla(k\nabla T) + \lambda_{vi}\nabla\left[\frac{\phi_g}{\tau}\frac{D_v M_w}{R_g T}\frac{P_{atm}}{P_{atm} - P_{ice}^*}\frac{dP_{ice}^*}{dT}\cdot\nabla T\right] - \lambda_f \rho_s\frac{dW_{l,eq}}{dT}\frac{dT}{dt} \tag{6.27}$$

or, after re-arrangement and putting $\rho_s = X_s\rho$:

$$\rho c_{app}\frac{\partial T}{\partial t} = \nabla\left[\left(k + \frac{\phi_g}{\tau}\frac{\lambda_{vi} D_v M_w}{R_g T}\frac{P_{atm}}{P_{atm} - P_{ice}^*}\frac{dP_{ice}^*}{dT}\right)\nabla T\right] \tag{6.28}$$

where

$$\rho c_{app} = \rho_{ds}\left[c_{ds} + \left(W_b + W_{l,eq}\right)c_l + W_{ice}c_{ice} + \lambda_f\frac{dW_{l,eq}}{dT}\right] \tag{6.29}$$

c_{app} is the apparent specific heat, dH/dT, which includes the latent heat of freezing (last term of Eq. 6.27). In Eq. 6.27 we have made use of the fact that the water vapour pressure in the food is that of ice when the latter is present in equilibrium.

In summary, the transport equations to be solved for heat and moisture transfer in a porous moist food can be written as follows:

Table 6.1 Parameters for heat and moisture transport (Eq. 6.30 and 6.31) in a moist porous food

If $(T > T_0)$ or $(W < W_{l,eq} + W_b)$, no ice is present:

$$\rho c_{app} = \rho_{ds}(c_{ds} + Wc_l) \text{ (Eq. 6.20)}$$

$$k_{eff} = k + \frac{\phi_g}{\tau} \frac{\lambda_{vl} D_v M_w a_w}{R_g T} \frac{P_{atm}}{P_{atm} - a_w P_{liq}^*} \frac{dP_{liq}^*}{dT} \text{ (from Eq. 6.21)}$$

$$D_{WT} = \frac{\phi_g}{\tau} \frac{M_w D_v a_w}{R_g T} \frac{P_{atm}}{P_{atm} - a_w P_{liq}^*} \frac{dP_{liq}^*}{dT} \text{ (from Eq. 6.16)}$$

If $(T \leq T_0)$ and $(W \geq W_{l,eq} + W_b)$, ice is present:

$$\rho c_{app} = \rho_{ds} \left[c_{ds} + (W_b + W_{l,eq})c_l + (W - W_b - W_{l,eq})c_{ice} + \lambda_f \frac{dW_{l,eq}}{dT} \right] \text{ (Eq. 6.29)}$$

$$k_{eff} = k + \frac{\phi_g}{\tau} \frac{\lambda_{vi} D_v M_w}{R_g T} \frac{P_{atm}}{P_{atm} - P_{ice}^*} \frac{dP_{ice}^*}{dT} \text{ (from Eq. 6.28)}$$

$$D_{WT} = \frac{\phi_g}{\tau} \frac{M_w D_v}{R_g T} \frac{P_{atm}}{P_{atm} - P_{ice}^*} \frac{dP_{ice}^*}{dT} \text{ (from Eq. 6.17)}$$

Equilibrium relationships (assuming Raoult's law):

$$W_{l,eq} = (W_0 - W_b) \frac{T_f - T_0}{T - T_0} \text{ (Eq. 6.25)}$$

$$a_w = \frac{P_{ice}^*(T_f)}{P_{liq}^*(T_f)} \text{ (Eq. 6.18)}$$

$$\rho c_{app} \frac{\partial T}{\partial t} = \nabla(k_{eff} \nabla T) \tag{6.30}$$

$$\rho_{ds} \frac{\partial W}{\partial t} = \nabla(D_{WT} \nabla T) \tag{6.31}$$

where the parameters c_{app}, k_{eff} and D_{WT} depend on the presence of ice and are given in Table 6.1. These transport equations can be solved by a numerical method using an explicit time stepping scheme, or an implicit iterative time stepping scheme.

There are two major sources of uncertainty in this model. One is the function $W_{l,eq}$, which determines the shape of the latent heat peak in the apparent specific heat curve c_{app}. Equation 6.25 is an approximation to Raoult's law. A more accurate expression will lead to an improved model. Another factor is the tortuosity factor. Because of the build up of ice and freezing expansion, the pores may gradually fill with ice causing τ to rapidly decrease. It is difficult to determine this effect without a detailed picture of the porous structure. This effect may be responsible for the build up of ice between crust and crumb when freezing part-baked bread, causing crust flaking (Le Bail et al. 2005; Hamdami et al. 2007).

6.2.3   Mass Transfer During Immersion Freezing

In immersion freezing the food is in contact with a liquid below its freezing temperature. The liquid may be a non-aqueous refrigerant or water containing an acceptable solute such as a salt, sugar, alcohol or a mixture of these, at a concentration high enough to maintain a liquid phase at low temperature. The use of ice slurry (Fikiin and Fikiin 1998) greatly enhances the process's efficiency by making use of the latent heat of the ice. If the food being frozen is not wrapped in an impervious film, solutes will diffuse into the food and water out of it, at the same time that heat is transferred. The objective may be to minimize this mass diffusion or to aim at some optimal value. For example, by freezing fruit in sugar-ethanol aqueous solutions or ice slurries, new dessert products can be formulated with beneficial effects on colour, flavour and texture, due to the enzyme-inhibiting action of the sugar (Fikiin et al. 2003). The objective of modelling in these cases is to predict both the temperature and the concentration profiles and histories in the food. As solutes penetrate the tissue, the freezing point will be depressed to various extents and it is important to take this effect into account.

Lucas et al. (2001) modelled the immersion freezing in a concentrated solution of a one-dimensional slab of inert porous material impregnated with dilute aqueous solution. The equations for diffusion of solute and for heat conduction were solved simultaneously. The porous medium was treated as a homogeneous phase, with average transport properties calculated from the fractions of ice, bead solid, water and solute. Diffusion takes place in the liquid channels only and hence the effective mass diffusivities must take into account the void fraction and tortuosity (Lucas et al. 1999). The transport equations were then solved by finite difference. The model was verified by the brine immersion freezing of a bed of glass bead impregnated with dilute NaCl solution. The model shows that at the start of the process, solute diffusion can prevent freezing at the surface and enhance the thawing of recently frozen layers. This results in a non-frozen surface layer co-existing with an inner frozen layer or frozen core. Continuing solute diffusion in the non-frozen layers causes the adjacent ice crystals to melt. If this process is allowed to continue, the product thaws completely and its solute concentration approximates that of the solution. A similar model was developed by Zorrilla and Rubiolo (2005) who used the enthalpy formulation for the heat transport equation.

6.2.4   Mass Transfer Between Intra- and Extracellular Spaces

Although neglected by food engineers, this topic is of potential importance in predicting the quality of foods, especially in conjunction with the modelling of intracellular nucleation. The phenomenon has been modelled in the field of cryosurgery and a numerical model, due to Devireddy et al. (2002), will be described later in the section on nucleation modelling.

6.3 Supercooling and Nucleation Effects

So far we have assumed that the freezing process is entirely governed by heat transfer. However, in many cases, the dynamics of nucleation and mass transfer have observable effects. It is well-known that foods almost never start to freeze at their thermodynamic freezing points. When water is cooled below the freezing point, it remains liquid until the temperature is low enough for stable ice crystals to form and grow. Above this nucleation temperature, any ice crystal that might form will lose molecules faster than it gains them due to surface energy effects (the curved surface of a crystal has higher free energy than a flat surface and therefore tends to lose molecules faster than it gains).

There are two types of nucleation: homogeneous and heterogeneous. Homogeneous nucleation happens only in pure water, in the absence of any foreign material, at a temperature of about $-40\,^{\circ}C$. In food freezing, heterogeneous nucleation is the prevailing mechanism except in the case of water in oil emulsions which will be considered later. It is caused by contact with a foreign material or with impurities, on which crystals form and grow. Heterogeneous nucleation happens at a higher temperature than homogeneous nucleation because the foreign material enables water molecules to form clusters on its surface with a large radius of curvature, thus lessening the surface energy.

If water or a water-rich material is cooled very quickly, nucleation does not have time to occur and the liquid in the food becomes an amorphous solid or glass, a process known as vitrification. Complete vitrification requires extremely fast cooling rates that are not likely to happen in ordinary food freezing. However, at high cooling rates there may be some partial vitrification, with ice crystal co-existing with a metastable solution that contains more water molecules than dictated by equilibrium due to the slow speed of diffusion. Over time, more ice may gradually crystallise out until equilibrium is attained, unless the material is below its glass transition temperature in which case diffusion processes become negligibly slow (Fig. 6.3). To date the effect of the metastable phase and glass transition on the freezing process has not been mathematically modelled, possibly due to the lack of reliable data. For example, the glass transition temperature T_g' for fully hydrated proteins has been variously reported as -20 to $-12\,^{\circ}C$ (for meat and fish) and -65 to $-85\,^{\circ}C$ (for beef, egg white, and tuna) (Simatos et al. 2009).

6.3.1 Instantaneous Nucleation Followed by Dendritic Crystal Growth

Some degree of supercooling is observed in most food freezing processes, where the surface dips briefly below the freezing point before suddenly coming up towards it. Nucleation usually happens very quickly, probably because the mechanical stresses caused by the appearance of the first crystals trigger further nucleation in the neighbourhood. Once nucleation has happened, the ice crystals will grow into the product, forming dendrites (see Sect. 6.4) and in most industrial freezing processes, the freezing process reverts to being heat transfer controlled. Pham (1989) modelled

Fig. 6.3 Concentration evolution of the non-ice phase during cooling on a binary phase diagram. *AB* slow equilibrium cooling, then past the eutectic point into a metastable region, then past the glass transition temperature T_g' into the glassy state at *B*; *CD* fast cooling through metastable region, followed by storage at constant temperature above glass transition during which ice continues to crystallise; *EF* fast cooling to a temperature below the glass transition

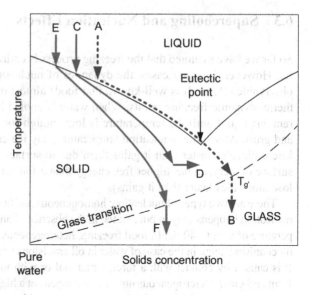

this type of behaviour with a finite difference model using the temperature correction method (Sect. 5.4.1.5) and validated the model with data from Menegalli and Calvelo (1978). Miyawaki et al. (1989) independently used the apparent specific heat technique to solve the same problem. The dendritic growth mechanism was verified by freezing layers of meat separated by plastic and measuring the temperature evolution (Menegalli and Calvelo 1978): the inner layer underwent its own nucleation, as evidenced by a second temperature jump.

To simulate supercooling, the specific heat (and thermal conductivity) above freezing is assumed to continue to apply below the initial freezing point, until the coldest node reaches nucleation temperature. At that point, the normal time stepping solution is momentarily stopped and all the nodes that have an enthalpy value H below freezing are assumed to freeze instantaneously, releasing enough latent heat for the node to warm up to the equilibrium temperature $T(H)$ (Fig. 6.4). Incidentally Fig. 6.4 resembles Fig. 5.9, which illustrates Pham's (1985) temperature correction step. Calculations continue normally throughout the material from that point onward. Pham (1989) found that, for the amount of supercooling that is commonly observed (a few degrees), supercooling has negligible effect on freezing time.

6.3.2 *Gradual Nucleation in an Emulsion*

When water is held as small droplets in an emulsion, such as in butter, each ice crystal cannot grow beyond its droplet and each droplet has to crystallize separately, a probabilistic phenomenon. In fact this effect is often used to study nucleation in the lab (Franks 2003). In such cases the freezing process may be very gradual and a freezing plateau may not even be present, as has been observed experimentally (Nahid et al. 2004, 2006, 2008).

Fig. 6.4 Modelling supercooling and nucleation on the
H-T diagram

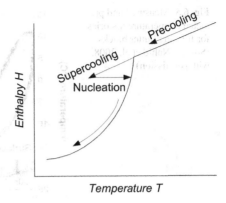

Nahid et al. used an explicit enthalpy method (Sect. 5.4.1.4) to solve the following heat conduction equation (obtained from Eq. 3.5):

$$\rho c \frac{\partial T}{\partial t} = \nabla \cdot (k\nabla T) + \rho L_f \frac{\partial F}{\partial t}$$

(6.32)

where F is the frozen fraction and depends both on the nucleation rate and the crystal growth rate (Avrami 1939, 1940, 1941). The nucleation rate was calculated by
Michelmore and Franks's (1982) equation, which predicts that the rate or nucleation
is a function of T/T_f. The crystal growth rate was assumed to be proportional to
$T_m - T$, where T_m is the equilibrium freezing point of the remaining solution (Müller
et al. 2004). This results in the following expression for the ice formation rate:

$$\frac{\partial F}{\partial t} = A(T_m - T)^{3/4} (V_{drop} e^{B\tau_\theta})^{1/4} (1 - F)[-\ln(1 - F)]^{3/4}$$

(6.33)

where τ_θ is a function of T/T_f, V_{drop} is the average droplet volume and A, B are
parameters from the nucleation and crystal growth models, which were found by
fitting an experimental temperature curve.

Using this model, Nahid et al. (2008) found that ice formation in butter can be de-
layed for days after the product has reached sub-freezing temperatures, causing large
temperature peaks to occur during cold storage. These results were confirmed by
experiments (Fig. 6.5) and have important practical implications for the heat load in
freezers, which will be overestimated, and cold stores, which will be underestimated.

6.3.3   Gradual Nucleation in Cellular Tissues

Another important reason for wanting to model nucleation and crystal growth is its
effect on food quality, cellular damage and drip loss. Maximum drip loss in meat
is believed to happen when a large intracellular crystal forms in each cell, which

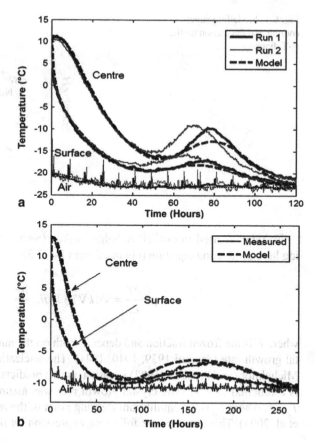

Fig. 6.5 Measured and predicted temperature histories for freezing butter blocks. (Source: Nahid et al. 2008, with permission)

causes maximal distortion and damage to the cell wall (Bevilacqua et al. 1979). This happens at an intermediate freezing rate, since faster freezing causes the formation of multiple small intracellular crystals, while slower freezing leads to extracellular freezing.

To date all the work on ice nucleation in cellular tissue has been carried out in the cryopreservation field. Irimia and Karlsson (2002) used a Markov chain model to calculate the nucleation and propagation of ice in cellular tissues and showed that ice can propagate from cell to cell. They also proposed (2005) a Monte Carlo technique to model the same problem.

Devireddy et al. (2002) developed a finite volume model to predict the formation of intracellular ice in biological tissues in the context of cryosurgery. The material was divided into two phases, extra- and intracellular. Extracellular liquid was assumed to freeze without supercooling. As it does so, the extracellular solute concentration increases, causing a difference in osmotic pressure, which results in the cell losing water at a rate:

$$\frac{dV_w}{dt} = -\frac{L_p A_{cell} R_g T}{v_w} \ln \frac{a_{w,in}}{a_{w,ex}} \tag{6.34}$$

where L_p is the cell membrane's permeability, A_{cell} the cell's surface area, v_w the molar volume of water, $a_{w,\,in}$ and $a_{w,\,ex}$ the water activities inside and outside the cell respectively. As a result the cell volume V_{cell} shrinks at a rate (Mazur 1963):

$$\frac{dV_{cell}}{dt} = -\frac{L_p A_{cell} R_g T}{v_w}\left\{\ln\left[\frac{V_{cell}-V_b}{V_{cell}-V_b+iv_w n_{salt}}\right]-\frac{M_w \lambda_f}{R_g}\left[\frac{1}{T_0}-\frac{1}{T}\right]\right\} \quad (6.35)$$

where i is the salt's dissociation constant, n_{salt} the number of moles of salt in the cell, V_b the volume of bound water in a cell and the other symbols are as previously defined. Due to this diffusion, intracellular solute concentration increases, but at a slower rate than extracellular fluid. However, the temperature in the cell falls just as fast as that of the extracellular medium, causing supercooling inside the cell. The probability of intracellular nucleation is obtained from the amount of supercooling, using a model by Toner et al. (1990, 1993).

To model the simultaneous thermal change and mass transfer, an iterative procedure has to be carried out at every time step to satisfy the heat balance as well as the intra- and extracellular mass balances:

Guess the rate of ice formation in each control volume;
Carry out the following iteration procedure:

  - Calculate the new nodal temperatures throughout the domain from the heat conduction equation (Eq. 3.5 or 6.1) and the amount of ice formed over the time step
  - Calculate the amount of extracellular ice from temperatures, assuming thermodynamic equilibrium in the extracellular space
  - Calculate the probabilities of intracellular nucleation, i.e. the increase in the number of cells with internal ice from Toner's nucleation model
  - Calculate the amount of intracellular ice
  - Calculate the total latent heat released by extra- and intra-cellular ice

until convergence is reached.

At every time step, material balances have to be set up to keep track of the amount of ice and unfrozen water inside and outside the cells. The model requires many parameters that are not commonly available about foodstuffs, such as cell size, cell surface area, cell membrane permeability, extracellular volume and parameters related to the onset of nucleation. As far as Bevilacqua et al.'s (1979) hypothesis about cell damage is concerned, it still does not allow the prediction of how many intracellular crystals will occur, and how large they will become.

So far, little work has been done in the food freezing field on the modelling of nucleation and crystal growth, although it seems to be a promising area of investigation. The main difficulty in applying these models is the difficulty of getting data on relevant parameters such as cell membrane mass transfer resistance and nucleation factors.

Fig. 6.6 *Top* dendritic
growth due to constitutional
supercooling, *bottom* profiles
of solute mole fraction x_{sol},
local freezing point T_f and
temperature. Growth velocity
(*grey arrows*) is proportional
to the driving force $T_f - T$,
hence growth is fastest at the
tip of the dendrite

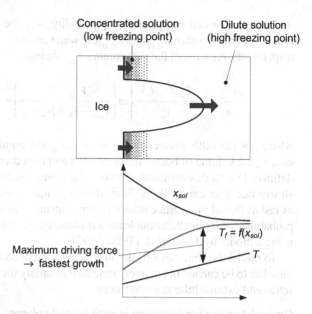

6.4   Microscale Modelling of Crystal Growth

In a liquid food, as the ice grows, solute is excluded from it and a concentration gradient will form in the liquid in front of it. This will increase the freezing point of the remaining water next to the freezing front (a phenomenon known as constitutional supercooling). If, due to irregularities in the front, part of the front protrudes into the unfrozen zone, it will be in contact with liquid at a lower concentration and hence higher freezing point. The driving force for crystal growth is $T_f - T$ (Müller et al. 2004), which is greater at the tip of the protrusion than at its base (Fig. 6.6). The protrusion will therefore grow faster than the rest of the front and form a dendrite. Further instabilities may cause dendrites to branch and form complex, snowflake-like patterns, trapping solutes in a solid matrix. At very fast freezing rates, solute diffusion may become significant in controlling the progress of the freezing front.

In the freezing or freeze concentration of liquids such as fruit juice, the rate of freezing affects the velocity of dendrite growth, which in turn influences the degree of separation of solute and ice. At low dendrite growth rate, solutes have time to diffuse away from the freezing front and there is good separation, but at high growth rate, solutes and suspensions become trapped in the ice matrix (Miyawaki et al. 2004). In solid foods, mass diffusion is usually very slow and therefore the solution becomes more and more concentrated locally as ice forms, lowering the freezing point and causing the characteristic gradual enthalpy-temperature curve. Dendrite growth may cause mechanical stresses and damages to cell membranes resulting in drip loss (Añon and Calvelo 1980). Recrystallization during frozen storage is an important factor leading to loss of sensory qualities (Martino and Zaritzky 1988, 1989). Modelling crystal growth in foods at the microscopic scale would provide

food engineers with valuable insight into these phenomena and enable them to optimise freezing processes, but to date it has not been widely done. Most of the work
has been on pure ice or in the metallurgical field (crystal growth in metals and
alloys), and more recently on biological cells in the context of cryobiology (e.g.
Jaeger et al. 1999; Jaeger and Carin 2002; Mao et al. 2003; Chang et al. 2007).

An approximate analytical equation for predicting crystal size from dendritic
growth theory was presented by Woinet et al. (1998a, b), assuming a Neumann
boundary condition, and validated against data from agar gel freezing. However,
precise predictions of crystal morphology require the use of numerical models.
A numerical model of crystal growth must take into account the following effects:

- Heat transfer.
- Mass diffusion of solutes if any.
- Convection, in the case of liquids.
- Curvature effect: surface curvature causes a depression of the freezing point, as
  given by the Gibbs-Thomson relationship

$$\frac{\Delta T_f}{T_f} = \frac{\sigma_{sl} K_R}{\rho_s \lambda_f} \qquad (6.36)$$

where σ_{sl} is the surface energy, T_f the freezing point for a flat interface, λ_f the latent
heat of freezing, and K_R the surface curvature ($K_R = 2/R$ for a sphere of radius R).

- Anisotropy: the atomic structure of crystals causes growth to be highly anisotropic.
- Kinetic effect: the freezing interface may not be at thermodynamic equilibrium
  due to the finite diffusion rate of atoms to the crystal surface.
- Random perturbations initiating the growth of dendrites.

Several approaches have been used to model crystal growth phenomena numerically: enthalpy methods, front tracking methods, level set methods, phase field
methods, and cellular automata.

6.4.1 Enthalpy Method

The enthalpy method presented in Sect. 5.4.1.4 can be adapted to calculate the
growth of crystals during solidification (Voller 2008; Karagadde et al. 2012; Pal
et al. 2013). A structured finite volume grid is used to discretize the domain. The
grid spacing must of course be much smaller than the size of the crystals. A control
volume is not allowed to commence the solidification process—i.e., will remain liquid, possibly supercooled—until at least one of the neighbouring control volumes
has completely solidified. The method is easy to program and fast, but the accuracy
of the predictions depends on the estimation of the local crystal surface curvature,
which has to be found by an approximate method since the front position is not
explicitly calculated.

6.4.2 Cellular Automata

In the cellular automata method (Rappaz and Gandin 1993; Brown 1998; Jarvis et al. 2000; Raabe 2001, 2002; Choudhury et al. 2012) the domain is also divided into very small control volumes or cells, each of which will switch states (from liquid to solid in this case) according to certain rules. The technique is similar to the enthalpy method described above once the heat conduction and mass diffusion equations are incorporated. However, cellular automata can be made to obey probabilistic rules, so they can also model nucleation, supercooling, anisotropy and dendritic branching quite naturally. Random perturbations arise spontaneously from the probabilistic rules and don't have to be artificially introduced as in deterministic methods.

6.4.3 Front Tracking or Sharp Interface Methods

Front tracking methods calculates the position and movement of the front at each instant directly, in parallel with solving the heat equation. The front speed v_f is given by the heat balance:

$$\rho \lambda_f v_f = k_f \left(\frac{\partial T}{\partial n} \right)_f - k_u \left(\frac{\partial T}{\partial n} \right)_u \qquad (6.37)$$

where n is the normal vector to the front and the subscripts f, u refer to the frozen and unfrozen phases respectively. Udaykumar et al. (1999) used a fixed cartesian finite volume grid and calculated the front position as follows: for each interfacial cell (a cell which contains the freezing front, i.e. with a frozen fraction greater than 0 and less than 1) the position of the point on the interface nearest the cell centre is calculated from the temperature field, then a piecewise quadratic curve isused to join the points on the front. Ye et al. (1999) and Udaykumar et al. (2002) used a piecewise linear curve to represent the front. Browne and Hunt (2004) represented the interface by marker points on two of the sides of each interfacial cell, and developed equations for the velocity of these markers.

6.4.4 Level Set Method

The level set method was devised by Osher and Sethian (1988) for computing and analyzing the motion of interfaces between phases, and introduced to the solidification field by Sethian and Strain (1992). In this method the interface position is not explicitly calculated (except for visualisation purposes) but is conceived as the contour $\varphi = 0$ of a field variable $\varphi(\mathbf{r}, t)$ at any instant. It can be shown that from a given initial front position, φ will evolve according to the partial differential equation:

Fig. 6.7 Interpretation of
the level set method. Over
time interval δt the variable
φ changes by $\delta\varphi$ causing the
interface to move by $v_f \cdot \delta t$.
It can readily be seen that in
the vicinity of the front
$\delta\varphi/(v_f \cdot \delta t)=-d\varphi/dn$, or
$d\varphi/dt+v_f \cdot d\varphi/dn=0$

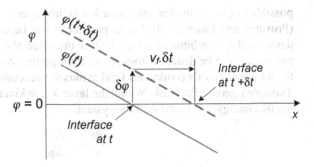

$$\frac{d\varphi}{dt} + F(\mathbf{r},t)|\nabla\varphi| = 0 \qquad (6.38)$$

where F is an artificially constructed smooth function that equals the front velocity
v_f at the interface. A graphical interpretation is shown in Fig. 6.7.

For numerical stability φ is usually taken to be the distance from the front
($|\nabla\varphi|=1$) (Chen et al. 1997), which is re-normalised after each time step by a
simple iterative procedure (Sussman et al. 1994). This results in a simplified form
of Eq. 6.38:

$$\frac{d\varphi}{dt} + F(\mathbf{r},t) = 0 \qquad (6.39)$$

Various approaches have been suggested for the construction of the extended velocity function $F(\mathbf{r}, t)$ (Adalsteinsson and Sethian 1999; Sethian 2001). A very simple
method was suggested by Tan and Zabaras (2006). The level set equation, Eq. 6.38,
is solved for a narrow band near the interface (Adalsteinsson and Sethian 1999)
while the heat and mass transport equations (and velocity field, in the case of a liquid medium) are solved for the whole domain. Because φ is a continuous function of
space, the interface curvature (at $\varphi=0$) can be easily and accurately calculated from
the derivatives of φ (Chen et al. 1997). Also, as the front position is not explicitly
tracked in the calculations, merging and breaking interfaces are easily and naturally
handled (Chen et al. 1997; Osher and Fedkiw 2002; Gibou et al. 2003). However, in
some versions of the level set method, the interface is still explicitly represented by
markers (as in the front tracking method) to allow accurate calculation of the front
velocity v_f (Sethian 2001).

6.4.5 Phase Field Method

In the phase field method (Boettinger et al. 2002; Sekerka 2004; Moelans et al.
2008; Provatas and Elder 2010), a phase field variable Φ is introduced. It takes
value 0 in the frozen phase (some workers use -1) and 1 in the liquid phase, and
varies smoothly from 0 to 1 over a layer of finite thickness δ_{pf} at the interface. It is

possible to give a physical interpretation to this layer in terms of mean field theory (Provatas and Elder 2010), but in practice it can be considered a computational device and δ_{pf} is arbitrarily selected. Once the value of δ_{pf} is chosen, all other model parameters can be deduced from physical properties. As with the level set method, the introduction of a continuous field avoids the necessity of tracking a sharp front, a complex geometrical task. Within the layer δ_{pf}, Φ kinetically evolves to minimise the free energy functional F of the system:

$$\frac{d\Phi}{dt} \propto -\frac{\delta F}{\delta \Phi} \tag{6.40}$$

where (ignoring anisotropy)

$$F = \int_V \left[f_\Phi(\Phi,T) + \frac{1}{2} a_\Phi |\nabla \Phi|^2 \right] dV \tag{6.41}$$

where a_Φ is a constant parameter and f_Φ is an energy function with local minima at $\Phi=0$ and $\Phi=1$. This results in a PDE for Φ which can be written in the form of a conventional transport equation:

$$\frac{d\Phi}{dt} = a_\Phi \nabla^2 \Phi + S_\Phi(\Phi,T) \tag{6.42}$$

where

$$S_\Phi = -\frac{\partial f_\Phi(\Phi,T)}{\partial \Phi} \tag{6.43}$$

A suitable form for f_Φ and hence S_Φ is (Kobayashi 1993; Wheeler et al. 1992):

$$f_\Phi(\Phi,T) = b\left[\frac{\Phi^4}{4} - \left(\frac{1}{2} - \frac{\beta}{3} \right)\Phi^3 + \left(\frac{1}{4} - \frac{\beta}{2} \right)\Phi^2 \right] \tag{6.44}$$

$$S_\Phi = b\Phi(1-\Phi)\left(\Phi - \frac{1}{2} + \beta \right) \tag{6.45}$$

where b is a constant related to the surface tension and β is a uniformly increasing function of $T - T_f$ which vanishes at the local freezing point T_f. The energy function f_Φ is plotted in Fig. 6.8. We notice that f_Φ is a potential well with two minima indicating stationary points at $\Phi=0$ and 1. The attraction range for $\Phi=0$ is greater than that of $\Phi=1$ at $T < T_f$ and less at $T > T_f$. A similar looking potential well was suggested by Caginalp (1986). Both formulations have been shown to converge to the sharp interface limit if the numerical parameters are suitably chosen (Caginalp 1989; Caginalp and Chen 1998; Wheeler et al. 1992; Fabbri and Voller 1997). The diffusive term in Eq. 6.42 can be regarded as a computational device to create a

Fig. 6.8 Plot of the free energy function f vs. phase field variable Φ at temperatures above ($\beta=0.2$), at ($\beta=0$) and below ($\beta=-0.2$) the freezing point

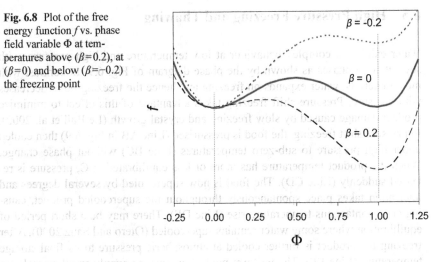

finite phase change layer; as a_Φ tends to 0 so does the layer's thickness (Kobayashi 1993). The phase field equation is solved over the layer δ_{pf} at the interface while the heat and mass transport equations (and velocity field, in the case of liquids) are solved for the whole domain.

Among the above methods the phase field method has been the most popular approach to the modelling of crystal growth for several reasons. In this model the surface energy is distributed over the diffuse layer and the Gibb-Thomson effect is automatically taken into account without the need to compute the surface curvature (Boettinger et al. 2002). Kinetic effects are also implicitly incorporated in the phase field equations. There is no need to accurately track the freezing interface which would require complex geometrical formulas.

Apart from some front tracking methods, the methods described in this section all use fixed grids, but adaptive grids, where the mesh is locally refined around the moving interface, are often necessary to reduce the computation time, especially in 3-D (Provatas et al. 1999). The simplest methods to implement are probably the enthalpy and cellular automata methods. These have not been popular for the prediction of crystal growth, perhaps because they do not give precise information on the freezing front position. However, this need not be a serious drawback, as the frozen fraction of each control volume can be calculated from its enthalpy (assuming we know if phase change has begun) and the front position can be recovered by a variety of geometric methods. For example, in the PLIC (piecewise linear interface construction) method, widely used in the computational fluid dynamics field, the interface is represented by a linear segment (for the 2-D case) in each partially frozen cell (control volume). Since the direction of the segment is normal to the (known) local temperature gradient, its precise location can be easily calculated from the frozen fraction of the cell (Gueyffier et al. 1999).

6.5 High Pressure Freezing and Thawing

Water exhibits a complex behaviour at low temperature and high pressure, with several solid phases as shown by the phase diagram of Fig. 6.9. In contrast with most materials, water expands on freezing and hence the freezing point decreases with pressure. Pressure shift freezing takes advantage of this effect to minimize product damage caused by slow freezing and crystal growth (Le Bail et al. 2002). In pressure shift freezing, the food is pressurised (Line AB in Fig. 6.9) then cooled under high pressure to sub-zero temperatures (Line BC) without phase change. When the product temperature has more or less equilibrated at C, pressure is released suddenly (Line CD). The food is now supercooled by several degrees and nucleation takes place spontaneously throughout the supercooled product, causing an instantaneous temperature rise (Line DE). There may be a short period of equilibration where some water remains supercooled (Otero and Sanz 2000). After freezing the product is further cooled at atmospheric pressure to its final storage temperature (Line EF). The uniform nucleation ensures evenly small crystal size and minimal textural damage.

High pressure thawing has also been investigated as a fast thawing method. Frozen food is pressurised (Line FG) then heated to its melting point (Line GH) where it gradually thaws. Once phase change is completed the food temperature rises (Line HI) and pressure is released (Line IA). Because of the lowering of the freezing point at H, the difference between product and ambient temperature is increased, hence a larger heat flux is obtained and thawing is accelerated. Also, microbial growth is less of a problem since lower ambient temperatures can be used.

In high pressure thawing and freezing, the effect of pressure on thermal properties (latent heat, freezing point, thermal conductivity) must be taken into account. Changes in calorimetric properties were considered in Sect. 2.4.7. Thermal conductivity below zero will also be different, since there is no ice.

Chourot et al. (1997) modelled high pressure thawing of an infinite cylinder of pure water, using FDM with Crank Nicholson stepping and the apparent specific heat approach. The latent heat is assumed to contribute a triangular peak spanning 1 K at the base. Thermal conductivity is assumed to be constant above and below the phase change range, and vary linearly over this range. The total latent heat and mean phase change temperature are given as polynomial functions of pressure. The entire thawing process takes place under pressure.

Denys et al. (1997) modelled pressure shift freezing using FDM with explicit stepping and apparent specific heat formulation. At the moment of pressure release, the temperature rise from T_i to T_{new} is calculated by an enthalpy balance. The product is assumed to be at uniform temperature when pressure is released; however, in a subsequent paper (Denys et al. 2000) this restriction is relaxed, and the energy balance is carried out node by node.

Pham (2006, 2008) suggested a flexible approach which can handle any temperature and pressure regime, using the enthalpy or temperature correction formulations, i.e. at every time step:

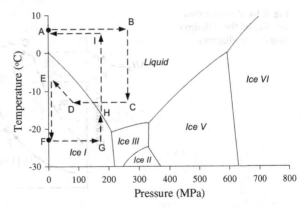

Fig. 6.9 Phase diagram of water at low temperature, with plotted paths of pressure-assisted freezing and thawing processes

- Calculate the nodal enthalpies from the enthalpy or temperature correction method
- Calculate the nodal temperatures from nodal enthalpies and pressure.

Norton et al. (2009) modelled the high pressure freezing by this method but also took into account the temperature and enthalpy changes due to compression and decompression, which are assumed to be adiabatic. They did this by calculating the nodal temperature change ΔT from the thermodynamic relationship

$$\Delta T_{adiabatic} = \frac{\overline{T}\beta_T}{\rho_u c_u}\Delta P \qquad (6.46)$$

then calculating the decompression enthalpy change from $c_u \Delta T_{adiabatic}$. However, this equation is unduly complicated and is problematic when the decompression is accompanied by phase change. A simpler thermodynamic expression is recommended:

$$\Delta H_{adiabatic} = \int \frac{dP}{\rho_u} \approx \frac{\Delta P}{\rho_{u,av}} \qquad (6.47)$$

Using this equation the effects of the compression-decompression steps are calculated by adding ΔH to the nodal enthalpies, just like the cooling steps, then temperatures are found from the H-T relationship. Figure 6.10 shows a pressure shift process on the enthalpy-temperature diagram. The food originally at state A is compressed to state B, cooled at high pressure from B to C, pressure is released causing nucleation at D (see also Fig. 6.9), then freezing is completed at atmospheric pressure along EF. From the programming point of view this presents no extra complication over a "standard" freezing program, apart from the need to construct an enthalpy curve for unfrozen food at high pressure. This has been described in Sect. 2.4.7.

In high pressure processing the product is usually in good thermal contact with a larger body of pressure-transmitting fluid, which also undergoes compressive

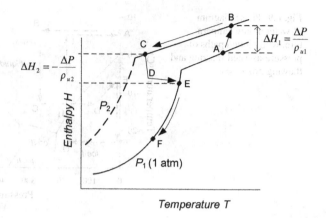

Fig. 6.10 Pressure shift freezing on the enthalpy-temperature diagram

Temperature T

heating and expansive cooling, therefore heat exchange with the fluid should be taken into account in a detailed model (Denys et al. 1997).

6.6 Freeze Concentration

6.6.1 General Principles

When an aqueous solution is frozen, pure ice crystals form leaving behind a more concentrated solution. This effect is applied in freeze concentration to obtain a concentrated liquid product without the risk of thermal damage. Two methods can be used: suspension freeze concentration, where a supercooled liquid is seeded with a suspension of ice crystals which gradually grow (Huige and Thijssen 1972), and layer freeze concentration, where ice grows on a cold surface (Müller and Sekoulov 1992; Flesland 1995; Miyawaki 2001; Miyawaki et al. 2005). To date, most or all industrial applications have used the former method. Mathematical models have been proposed to predict the rate of ice growth and the distribution of solutes between the ice and liquid phases.

In equilibrium freezing the ice is pure H_2O, but in practice some solute and suspended solids may be trapped within the ice matrix, reducing the separation efficiency and causing wastage. The ratio of solutes concentration in the frozen phase to that in the liquid phase is the distribution coefficient, K_d, which must be minimised:

$$K_d \equiv \frac{X_{sol}^{ice}}{X_{sol}^{liq}} \tag{6.48}$$

The distribution coefficient depends on the crystal growth rate. Due to the separation of solids from the ice, concentration and freezing point profiles arise in the

Fig. 6.11 Temperature, concentration and freezing point profiles near a growing ice front

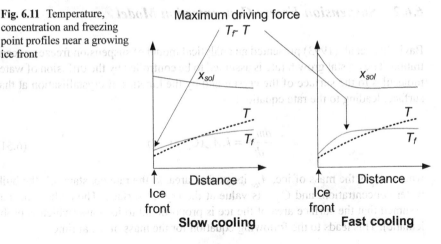

solution. At low freezing rates, these profiles are flat because solutes have time to diffuse away into the bulk of the liquid, and the driving force for crystallization, $T_f - T$, is maximal at the ice surface, resulting in planar ice growth (Fig. 6.11). At high freezing rates, the concentration and freezing point profiles are steep, resulting in the maximum driving force being located away from the ice surface. This is known as constitutional supercooling and results in dendritic growth, which tend to trap solutes and other solids in the ice matrix (Fig. 6.6). Thus a criterion for stable and pure crystal growth is that at the ice front we must have (Rutter and Chalmers 1953):

$$\left(\frac{dT_f}{dx}\right)_{liq} \leq \left(\frac{dT}{dx}\right)_{liq} \tag{6.49}$$

Interestingly, this suggests that in layer freeze concentration one can use a faster freezing rate with a hotter solution without compromising solute separation. The condition of Eq. 6.49 is never obtained in suspension freeze concentration, because the solution is supercooled and therefore $dT/dx < 0$ while $dT_f/dx > 0$, but perhaps trapping of solids is less likely on a spherical surface. In practice, the distribution coefficient is determined experimentally as a function of the ice front velocity (Miyawaki et al. 2005; Gu et al. 2005). For layer freeze concentration, Chen and Chen (2000) proposed the following equation, which agrees with data for a variety of solutions and emulsions (salt, sugars, orange juice, milk):

$$K_{d,av} = -0.10 + 0.32(T_0 - T_f) - 0.04(T_0 - T_f)^2 + 0.12\frac{v_{f,av}}{v_\infty^{0.5}} \tag{6.50}$$

where $T_0 - T_f$ is the freezing point depression, $v_{f,av}$ the average velocity of the ice front and v_∞ the bulk velocity of the solution.

6.6.2   Suspension Freeze Concentration Model

Bayindirly et al. (1993) presented an analytical model of suspension freeze concentration. The crystal growth rate is assumed to be controlled by the diffusion of water molecules to the surface of the crystal and by the kinetics of crystallisation at that surface, leading to the rate equation:

$$\frac{dm_{ice}}{dt} = k_1 A_{ice}(C_w - C_{w,s})$$ (6.51)

where m_{ice} is the mass of ice, A_{ice} its surface area, k_1 the rate constant, C_w the bulk water concentration, and $C_{w,s}$ its value at the crystal surface. The authors further assumed that the surface area of the ice is proportional to its mass (which is problematic). This leads to the following equation for the mass of ice at time t:

$$m_{ice} = \frac{m_{ice}^0 e^{\xi t}}{1 - \frac{m_{ice}^0}{m_{ice}^{max}}(1 - e^{\xi t})}$$ (6.52)

where ξ is a process parameter and superscripts 0 and max refer to the initial and maximum (final) masses of ice respectively. This equation (known as a logistic function) shows that the mass of ice increases slowly at first (due to the limited surface area for crystal growth), then more rapidly as more area becomes available, then slows down and approaches an asymptotic maximum value as the water concentration approaches saturation. The concentration of the juice can be found by material balance, assuming the distribution coefficient is known.

6.6.3   Layer Freeze Concentration Models

Ratkje and Flesland (1995) proposed a model for a falling film freezing on a cold surface (Fig. 6.12) based on irreversible thermodynamics. The ice front velocity v_f is assumed to be limited by the diffusion flux of water molecules to the surface, J_w, which depends on both the water concentration gradient and the temperature gradient in the liquid:

$$v_f \propto J_w \propto \frac{M_w \lambda_f}{T}\left(\frac{dT}{dx}\right)_{liq} - \frac{R_g T}{C_w}\left(\frac{dC_{sol}}{dx}\right)_{liq}$$ (6.53)

Unfortunately there appears to be an error in the coefficient of proportionality in the paper (Eq. 30), which results in the predicted ice growth rate becoming infinite when the volumetric expansion upon freezing is ignored. Chen et al. (1997) pointed

Fig. 6.12 Temperature and
concentration profile in layer
freeze concentration

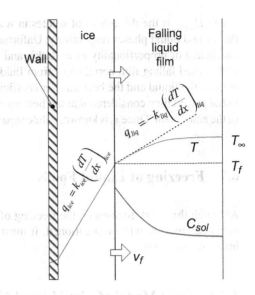

out that the above equation predicts that the ice growth rate would increase with temperature gradient in the fluid, i.e. when the fluid gets hotter, a paradoxical result. However, the model implicitly assumes that the cooling rate is sufficient to carry away, through the ice layer, all the latent heat of freezing and maintain the interface at or below the freezing point. Thus, ultimately the controlling factor for ice growth in this model is still heat transfer.

Most other models for layer freeze concentration are based on the heat balance at the ice-liquid interface, which can be written as:

Latent heat release = heat leaving through ice—heat arriving from liquid

$$\rho_{ice}\lambda_f v_f = k_{ice}\left(\frac{dT}{dx}\right)_{ice} - k_{liq}\left(\frac{dT}{dx}\right)_{liq} \qquad (6.54)$$

or in terms of a heat transfer coefficient h at the ice-liquid interface:

$$\rho_{ice}\lambda_f v_f = k_{ice}\left(\frac{dT}{dx}\right)_{ice} - h(T_\infty - T_f) \qquad (6.55)$$

where T_∞ is the bulk liquid temperature and $(dT/dx)_{liq}$ or h are calculated from fluid dynamics relationships (Fig. 6.12). Models of this type were presented by Flesland (1995), Chen et al. (1997), Audela et al. (2011) and others. Chen et al. (1997) added to the right hand side of Eq. 6.54 the sensible heat carried by the diffusion fluxes of water and solute, obtaining:

$$v_f \propto k_{ice}\left(\frac{dT}{dx}\right)_{ice} - k_{liq}\left(\frac{dT}{dx}\right)_{liq} - D_{sol,w}M_{sol}c_{sol}T\left(\frac{dC_{sol}}{dx}\right)_{liq} - D_{sol,w}M_w c_w T\left(\frac{dC_w}{dx}\right)_{liq}$$

$$(6.56)$$

where $D_{sol,\,w}$ is the diffusivity of solutes in water and subscripts *ice* and *liq* refer to the ice and liquid phases respectively. Unfortunately the same error was made in the coefficient of proportionality as in Ratkje and Flesland's (1995) paper. Audela et al. (2011) used falling film correlations from fluid dynamics theory to calculate the velocity of the liquid and the heat transfer coefficient at the solid-liquid interface. In the initial period they considered that the heat transfer is limited by the cooling capacity of the equipment. Once v_f is known, solute separation can be calculated from Eq. 6.50.

6.7 Freezing of Liquid Foods

Although this work focuses on the freezing of solid foods, a few pertinent features of liquid freezing will be mentioned. It must be said that this topic has not been investigated widely.

6.7.1 CFD Model of Liquid Food Freezing

Computational fluid dynamics (CFD) techniques must be used to model fluid motion and convective effects in a liquid. The heat transport equation now incorporates a convection term $\rho c \mathbf{v} \cdot \nabla T$ where $\mathbf{v}$ is the velocity vector of the fluid, and becomes:

$$\rho c \frac{\partial T}{\partial t} + \rho c \mathbf{v} \cdot \nabla T = \nabla \cdot (k \nabla T) + S_q \qquad (6.57)$$

while additional transport equations must be solved for continuity and momentum transport (one for each direction of space), and, if the flow is turbulent, for the variables characterising the turbulence field (such as turbulence intensity k and turbulence dissipation ε in the k-ε model). The energy equation can be solved using FVM or FEM (Chap. 5).

The main problem with the momentum equation is how to deal with the solidification of the fluid, which causes velocity to vanish. The momentum equation can be written as:

$$\rho \frac{\partial \mathbf{v}}{\partial t} + \rho \mathbf{v} \cdot \nabla \mathbf{v} = \mu_{eff} \nabla^2 \mathbf{v} - \nabla P + \mathbf{S}_v \qquad (6.58)$$

where μ_{eff} is the effective viscosity (including turbulent viscosity) and $\mathbf{S}_v$ is the momentum source term (usually the gravitational force). Bennon and Incropera (1987) reviewed the various methods of handling solidification. In one class of methods, the solid and liquid phases are treated as different domains with a clear boundary between them (similarly to the moving grid method for solving the freezing of solid foods). The method is complicated and cannot be applied to materials which do not exhibit a sharp phase change temperature.

A second approach involves changing the viscosity. The viscosity of the fluid is specified as a function of temperature and increases more or less gradually by several orders of magnitudes as the temperature falls below the freezing point. A reasonable freezing temperature range must be defined to avoid too sharp a change, which would cause convergence problem. The method is easy to implement (commercial numerical software normally allow viscosity to be entered as a function of temperature) and intuitively attractive, since it reflects the behaviour of most real materials. However, if a realistic viscosity-temperature curve is used, convergence will be very slow or impossible due to the very large and steep change in viscosity over a small temperature range. Therefore a much smaller value of solid viscosity may have to be used, and the solution will only be approximate, with a small but finite residual velocity in the frozen region.

A third method termed the *source method*, or *porosity method*, is employed in commercial CFD software packages such as CFX and Fluent, which are extensively used to model the casting of metals and polymers. From the temperature field, the frozen fraction at each point is calculated. The frozen fraction is visualised as a porous solid phase, which causes a resistance to the flow of the liquid fraction. This resistance is entered into the momentum equation as a negative source term, which is proportional to velocity and increases with the frozen fraction F according to the Kozeny-Carman relationship:

$$S_v = -A\frac{F^2}{(1-F)^3}\mathbf{v} \qquad (6.59)$$

A is a proportionality constant which is related to the viscosity and drag coefficient. A large value of A causes a sharper transition from liquid to solid behaviour, but may cause convergence problems. A small number is usually added to the denominator of Eq. 6.59 to prevent the source term from going to infinity.

Most food-related CFD modelling works use commercial software. Norton and Sun (2013) gave a review of recent applications and list commonly used packages. There have been few applications of CFD to the freezing of liquid foods. Lian et al. (2006) used the commercial CFD code FIDAP to model fluid flow and heat transfer in a scraped surface freezer. From the temperature and velocity profiles, the rates of ice nucleation and crystal growth can be calculated and substituted into a population balance equation, which keeps track of the number of particles in each size range:

$$\frac{\partial n}{\partial t} + \mathbf{v}\cdot\nabla n + \frac{\partial(Gn)}{\partial R_p} = B\delta(R_p - R_{p0}) \qquad (6.60)$$

where n is the local population density of crystals of size R_p, $\mathbf{v}$ is the fluid velocity, G the crystal growth rate, and the right hand side is the rate of nucleation (which is zero for all particle sizes except the minimum crystal size R_{p0}). The nucleation rate B and growth rates G are both function of local deviations from equilibrium concentration. The model was used to predict the mass fraction, number density and mean size of ice crystals during the crystallisation of a sucrose solution. However, viscosity was assumed to be constant, which is unrealistic.

6.7.2   Freezing of a Well-Stirred Liquid

Consider a liquid contained in a container in an environment at temperature $T_a < T_f$. If the liquid is perfectly stirred, its temperature will be uniform and the cooling rate before freezing (the precooling period) will be given by the heat balance:

$$\rho c V \frac{dT}{dt} = hA(T - T_a), \; T \geq T_f \qquad (6.61)$$

where h is the heat transfer coefficient between the fluid and the environment, leading to the predicted value of the precooling time:

$$t_{precool} = \frac{\rho c V}{hA} \ln \frac{T_i - T_a}{T_f - T_a} \qquad (6.62)$$

When $T = T_f$ a layer of ice forms on the boundaries and gradually thickens. Assume for the moment that the food is an infinite slab with thickness $2R$ and that the ice-liquid interface is at distance x_f from the centre, then the thickness of the freezing front will be $R - x_f$. If the heat transfer coefficient between the wall and the ice is h_s and that between the ice and the liquid is h_l, the heat flux becomes:

$$q = \frac{T_f - T_a}{\dfrac{1}{h_s} + \dfrac{R - x_f}{k_f} + \dfrac{1}{h_l}} \qquad (6.63)$$

where $(R - x_f)/k_f$ is the thermal resistance of the ice layer. We now consider some special cases:

a. If T_f is constant (pure water) and the geometry is that of a slab, this equation becomes identical to Eq. 3.10 from which Plank's equation was derived as long as $1/h$ is replace by $1/h_s + 1/h_p$, therefore Plank's equation for the slab is also applicable to the phase change period of this case, provided the zero sensible heat assumption is maintained. The approximate and empirical equations *for a slab* developed in Chap. 4 can also be used to calculate the duration of the phase change-subcooling period, with all variables as defined in Sect. 4.2.4. For shapes other than a slab, the shape factors developed in those chapters are not necessarily applicable, since the heat transfer coefficient h_l applies at the liquid-ice interface, which is continually shrinking.

b. If T_f is constant and $h_l R/k_f >> 1$ (which is usually the case when the stirring is vigorous) then the term $1/h_l$ in Eq. 6.63 can be ignored, which leads back to Plank's equation. In this case all the analytical, approximate and empirical formulas developed for the freezing of a solid with infinite Bi will also be applicable to *all* shapes to calculate the phase change-subcooling period, with similar reliability, provided h is replaced by h_s.

c. If the liquid contains solutes, their concentrations will increase with time as ice separates out and T_f will gradually fall, so the formulas for the freezing of solids cannot be used. This effect has been explored in Section 6.6 (Freeze Concentration).

6.8 Microwave and Radio Frequency Thawing

6.8.1 General Principles

The use of dielectric heating (microwave and radio frequency, or RF) can greatly accelerate the thawing process. Microwave and RF are both part of the electromagnetic spectrum, with RF having frequencies below 300 MHz while microwaves have frequencies from 300 to 3000 MHz (these boundaries can be flexible). In RF heating an alternating electric field is generated by plate electrodes on each side of the product, while in microwave heating microwave travels from a separate generator to the product enclosure via a waveguide. Because of its lower frequency and longer wavelength, RF has greater penetration depths (typically tens of cm for RF compared with a few cm for microwave) and can give more uniform heating, but the equipment is more costly. With microwave thawing there is a great risk of uneven absorption and runaway heating, because of higher absorption of microwave in liquid water compared to ice. Any part of the product that thaws first will continue to heat up more quickly and may become cooked before thawing is complete in the rest of the food. This may be remedied by using an intermittent microwave source to give time for the temperature profile to become more uniform. Microwave tempering, where the product is brought to a temperature just below melting point in order to soften it for cutting and slicing or before completing the thawing process by conventional means, is much easier to control and hence widely used. An important objective of the modelling of microwave thawing is therefore to predict the occurrence of local heating.

The heat transport equation can be solved numerically. The key question is how to model the source term S_q to accurately represent the heat generated by the microwaves. There are two approaches (Ayappa et al. 1991; Budd and Hill 2011): Lambert's Law and Maxwell's equations.

6.8.1.1 Lambert's Law

Lambert's law states that equal thicknesses of a material absorb equal fractions of incident power, and therefore the power flux I varies exponentially with distance x:

$$I(x) = I_0 e^{-2\beta x} \tag{6.64}$$

where I_0 is the flux at the surface and β is the attenuation coefficient. The inverse of 2β is called the penetration depth. The power absorbed per unit volume is therefore:

$$-\frac{dI(x)}{dx} = 2\beta I_0 e^{-2\beta x} \tag{6.65}$$

Lambert's law is valid for a semi-infinite solid where there is little reflection on the opposite surface, and is a good approximation when the thickness L satisfies the criterion (Ayappa et al. 1991)

$$L \geq 2.7\beta^{-1} - L_0 \tag{6.66}$$

where $L_0 = 0.0008$ m. The critical size is several times greater for a cylinder (Oliveira and Franca 2002). When the above criterion is not satisfied, Maxwell's equations must be used.

6.8.1.2 Maxwell's Equations

Maxwell's equations describe the interactions between the electric and magnetic fields in space (Feynman 2014). Food is a non-magnetic, electrically neutral medium and may be considered non-conducting in a microwave or RF field. For such a material, Maxwell's equation leads to (Zhang and Datta 2000):

$$\nabla\left(\frac{\nabla\varepsilon}{\varepsilon} \cdot \mathbf{E}\right) + \nabla^2\mathbf{E} - \mu_0\varepsilon\frac{\partial^2\mathbf{E}}{\partial t^2} = 0 \tag{6.67}$$

where $\mathbf{E}$ is the electric field μ_0 the permeability of free space and ε the permittivity of the medium. For an electromagnetic wave with angular frequency ω this leads to:

$$\nabla\left(\frac{\nabla\varepsilon}{\varepsilon} \cdot \mathbf{E}\right) + \nabla^2\mathbf{E} + \mu_0\varepsilon\omega^2\mathbf{E} = 0 \tag{6.68}$$

ω is a complex angular frequency whose real and imaginary parts depend on the dielectric properties ε' and ε'' (see below) of the medium. Zhang and Datta (2000) stressed that the first term in Eq. 6.68, which appears in a non-homogeneous medium, affects the fields non-linearly and is caused by the permittivity gradient, which is significant in temperature sensitive materials such as foods (and especially so in thawing). However, in some works this term is ignored, leading to the homogeneous media equation:

$$\nabla^2\mathbf{E} + \mu_0\varepsilon\omega^2\mathbf{E} = 0 \tag{6.69}$$

For a non-conducting material the complex permittivity ε is usually written as:

$$\varepsilon = \varepsilon' + i\varepsilon'' = \varepsilon_0 (\kappa' + i\kappa'') \qquad (6.70)$$

where ε_0 is the permittivity of free space. ε' (or κ') measures the material's ability to store electrical energy and ε'' (or κ'') its capacity for energy dissipation. The power dissipation per unit volume, i.e. the microwave heat source term, is given by (Ayappa et al. 1991):

$$S_q = \frac{1}{2} \omega \varepsilon'' |\mathbf{E}|^2 \qquad (6.71)$$

During food thermal processing and especially thawing, food dielectric properties change strongly with temperature (at 2.45 GHz microwave penetration depth is almost three orders of magnitude higher in ice than in water) and therefore the electromagnetic and heat transfer equation are very strongly coupled. A realistic model must take this coupling into account. The property parameters ε' and ε'' (or κ' and κ'') for many foods have been measured at several frequencies and listed by Ayappa et al. (1991). A recent review of data can be found in Sola-Morales et al. (2010). Prediction methods for foods were recently reviewed by Gulati and Datta (2013).

6.8.2 Analytical Solutions

Ayappa et al. (1991, 1997) gave analytical solutions of Eq. 6.69 for a multilayer slab and a cylinder respectively for both Lambert's law and Maxwell's equation. Lambert's law predicts that power absorption falls smoothly with distance from the surface, while Maxwell's equation predicts a sinusoidal pattern superimposed on the decreasing trend with distance.

6.8.3 Numerical Solutions

Oliveira and Franca (2002) carried out finite element calculations for the heating of slabs, cylinders and foods of irregular shapes. Maxwell's equations was discretized and solved along with the heat conduction equation. Changes of dielectric properties with temperature were not taken into account, i.e. Eq. 6.69 was used. Equation 6.71 was used to calculate the heat generation. Good agreement with the analytical solution for a slab was obtained. Oliveira and Franca also simulated the effect of product rotation and intermittent microwave power and found that, when applied in combination, they gave more uniform heating than each separately.

For more realistic predictions, the effect of temperature on dielectric properties and the effect of the design of the microwave equipment and placement of the product in the cavity on the electromagnetic field must be taken into account. Zhang and Datta (2000) built such a model using two independent commercial FEM packages, EMAS for electromagnetic field calculations and NASTRAN for heat conduction in a solid product. The two software packages were coupled at the computer's operating system level. First the electromagnetic field is calculated (Eq. 6.67), then the absorbed power during a time interval (Eq. 6.71). The temperature profile at that point in time is then calculated by solving the heat conduction equation in NASTRAN, then the dielectric properties ε' and ε'' are calculated for each node and fed to EMAS to compute the magnetic field in the next time step.

Newer FEM packages have multiphysics capability, which means that they can solve several field equations simultaneously so that manual coupling of different software is not necessary. Salvi (2011) compared two multiphysics FEM packages, Comsol and Ansys, on a 3-D problem involving the continuous flow microwave heating of a liquid (fresh water and 1.5% salt solution), a fairly demanding problem that involves the solution of Maxwell's equation, heat transfer and fluid flow. For electromagnetic problems, the element size L_{el} requirement was decided based on the Nyquist criterion $L_{el} < \lambda/2$ where λ is the wavelength. Results of the comparison were:

- Agreement of predictions: the power absorption predicted by Comsol was up to 15% higher than that by Ansys. Predicted temperature profiles were qualitatively similar (same hot spot location) but there were discrepancies of up to 25 K for the maximum temperature in the salt solution. The predicted average temperature rise was the same for both packages.
- Ease of use: Comsol was found much easier to use. Ansys required the user to specify two separate grids, one for electromagnetism and another for fluid and heat transport, and the results for each were exported to a separate file.
- Coupling of equations: in Comsol the electromagnetism equations were solved first then heat and fluid transport. Comsol uses more memory and full coupling of the equations was not possible with the computer available (3 GHz Xeon processor with 3 GB RAM). In Ansys full coupling was carried out with the three equations solved iteratively, resulting in possibly more accurate solution.
- Computation time: 15 min for Comsol, two hours for Ansys (due to the one-way coupling of physics in Comsol and full coupling in Ansys).

6.9 Thermomechanical Effects During Freezing

Water expands by about 9% by volume when turning into ice, causing considerable stresses in foods during freezing. In cryogenic freezing, this expansion is followed by a significant thermal contraction, of the order of 0.5% in linear terms or 1.5% in volumetric terms (Rabin et al. 1998). Frozen food is brittle and these stresses may cause cracking in the food, especially at high cooling rates such as in cryogenic freezing. Rubinsky et al. (1980) carried out an approximate analytical analysis of thermal stresses during the freezing of organs but neglected phase-related volume change. Rabin and Steif (1998) calculated thermal stresses in freezing a sphere, taking both phase change expansion and thermal contraction into account, but assumed that the unfrozen material is liquid and neglected the property changes due to freezing. Shi et al. (1998, 1999) carried out thermal strain and stress calculations using the commercial software ABAQUS (ABAQUS Inc., Rhode Island, USA). They used both an elastic model and a viscoelastic model, but neglected thermal contraction. Tremeac et al. (2007) used the same software to simulate thermal stresses during the freezing of a two-layered food.

At moderate values of strains and stresses, thermophysical properties can be assumed to be unaffected; hence the analysis can be carried out in two stages: the thermal history is calculated first, using any of the methods listed earlier, followed by stress and strain calculation. This procedure is not necessarily valid when stress values are very large, such as when the volume is constrained, generating very high pressures and consequent thermal property changes. The stress analysis assumes that total strain is the sum of thermal strain $\varepsilon_{ij}^{(T)}$ (due to temperature change) and mechanical strain $\varepsilon_{ij}^{(m)}$ (due to mechanical stresses):

$$\varepsilon_{ij} = \varepsilon_{ij}^{(m)} + \varepsilon_{ij}^{(T)} \tag{6.72}$$

where $i, j = 1, 2$ or 3 are space coordinate indices. The thermal strains can be calculated as a function temperature and should include both the phase change expansion and thermal contraction of ice:

$$\varepsilon_{ij}^{(T)} = \delta_{ij} \int_{T_{REF}}^{T} \beta_T \, dT \tag{6.73}$$

To solve for the mechanical strains, some constitutive relationships for the material must be assumed. In an elastic model, strains are linearly related to the (present) stresses:

$$\sigma_{ij} = \delta_{ij} K e_{kk}^{(m)} + G e_{ij}^{(m)} \tag{6.74}$$

while in a viscoelastic model, strains depend on stress history and vice versa:

$$\sigma_{ij} = \delta_{ij} \int_0^t K(t-\tau) \frac{\partial e_{kk}^{(m)}}{\partial \tau} d\tau + \int_0^t G(t-\tau) \frac{\partial e_{ij}^{(m)}}{\partial \tau} d\tau \tag{6.75}$$

where e_{ij} are the deviatoric strains, defined by

$$e_{ij} = \varepsilon_{ij} - \delta_{ij} \frac{\sum_k \varepsilon_{kk}}{3} \tag{6.76}$$

The mechanical stress thus consists of an expansion component and a shear or deviatoric component. K is the bulk modulus ($K \equiv \rho d\rho/dP$) and G the shear modulus of the material, i.e. the stresses caused by a unit step in the strain. For elastic materials, both K and G are constant, while for viscoelastic materials they are functions of time. The integrals in Eq. 6.75 are termed "hereditary integrals". They are obtained by assuming that an arbitrary strain pattern can be decomposed into a series of steps $\Delta\varepsilon$ each happening at some previous time τ, and the stress at time t resulting from each step, $\Delta\varepsilon \cdot K(t-\tau)$ and $\Delta\varepsilon \cdot G(t-\tau)$, can be summed up or integrated.

The various components of stress and strain are not independent but are related by equilibrium relationships:

$$\sum_j \frac{\partial \sigma_{ij}}{\partial x_j} + F_i = 0 \tag{6.77}$$

where F_i is the i-th component of body force, and compatibility conditions arising from degrees of freedom considerations:

$$\frac{\partial^2 \varepsilon_{ij}}{\partial x_k \partial x_l} + \frac{\partial^2 \varepsilon_{kl}}{\partial x_i \partial x_j} - \frac{\partial^2 \varepsilon_{lj}}{\partial x_k \partial x_i} - \frac{\partial^2 \varepsilon_{ki}}{\partial x_l \partial x_j} = 0 \tag{6.78}$$

In the case of spherical and cylindrical foods, the problem is greatly simplified by the disappearance of most of the terms in the stress and strain equations. In spherical coordinates, for example, most of the terms in the stress tensor disappear, leaving only two: the radial normal stress σ_r and the tangential (or circumferential, or azimuthal) stress σ_t. The compatibility equations reduce to:

$$\varepsilon_r = \frac{du}{dr} \tag{6.79}$$

$$\varepsilon_t = \frac{u}{r} \tag{6.80}$$

where u is the radial displacement. By considering the forces acting on a thin shell, the equilibrium relationships reduce to:

$$\frac{d\sigma_r}{dr} + \frac{2}{r}(\sigma_r - \sigma_t) = 0 \tag{6.81}$$

and the system of equations can easily be solved once the constitutive relationships and thermal fields are known.

An important consideration is the material's hydrodynamic behaviour at the freezing front. Two different models have been proposed, with very different predictions of strains and stresses, depending on what happens as the material changes phase and expands (McKellar et al. 2009). In the first, termed the isotropic expansion model, the material is considered as a solid both before and after freezing and expands isotropically upon phase change, except for the effect of stresses in the different directions (Shi et al. 1998; Pham et al. 2006b, c). At a microscopic level, this model assumes that the water molecules locally re-arrange themselves into crystals without otherwise moving with respect to each other. In the second model, termed the isotropic stress model, water molecules move freely out of the freezing front as its density reduces, maintaining a state of zero shear stress at the front (Rabin and Steif 1998). The first model predicts that tangential stress in the frozen shell is compressive, until freezing is completed when all stresses vanish. Freeze cracking is therefore unlikely except at cryogenic temperatures, when the outer shell undergoes thermal shrinkage. The second model predicts high tensile tangential stress in the outer shell and high compressive stresses in the core, irrespective of the environmental temperature (as long as freezing occurs). By measuring the rate of expansion of potatoes during freezing, McKellar et al. (2009) found that the isotropic expansion model fits the data better for this material. However, it is speculated that other materials may behave differently, depending on whether they are more liquid-like or more solid-like in the unfrozen state.

Chapter 7
Conclusions

The numerical modelling of the classical "pure thermal" freezing problem can be considered solved in principle. An enthalpy or quasi-enthalpy method is recommended, in conjunction with control-volume FDM, lumped capacitance FEM or FVM. Explicit time stepping is recommended for small or one-off problems, Pham's quasi-enthalpy method for those who want speed as well as uncomplicated programming. Iterative enthalpy methods are useful from a mathematician's point of view to provide rigorous second-order results which guarantee strict energy balance at all time steps. However, if commercial software such as Comsol, ABAQUS, Fluent or CFX is used, it may be difficult for the user to apply the enthalpy or quasi-enthalpy methods as they are not a standard option, or may not be an option at all. The (iterative) effective specific heat method seems to be the only practical approach with these packages. To obtain reasonable accuracy and computing speed, the latent heat peak may have to be "smeared out" and have reasonable width. The FEM software Ansys (2005) seems to be the exception as it has an iterative enthalpy method option for nonlinear thermal problems and also allows the choice of diagonal heat capacity matrix (i.e. lumped capacitances).

In spite of advances in calculation methods, significant errors may still arise in their application to real-life problems, due to uncertainties in the inputs. Further research will therefore need to concentrate on the following factors:

1. The surface heat transfer coefficient is not always accurately known. It is difficult to measure directly and often has to be calculated from other factors using empirical formulas: air or fluid velocity, product geometry, air or fluid physical properties, radiation temperatures, etc. Condensation, frosting or evaporation of water will affect its value significantly. These effects often vary during the freezing or thawing process due to changes in product temperature. The effect of heat transfer coefficient is more pronounced at small Biot numbers (see Sect. 3.2.3).

2. Thermal properties are not always accurately known. Foods are natural products or made from natural ingredients, and their composition varies unpredictably. No fish or chicken is identical to another. Even if the composition is accurately known, there are still errors involved in the methods used to calculate thermal conductivity, ice content and calorimetric properties. For example, thermal conductivity depends on the ice microstructure, which will differ depending on

Q. T. Pham, *Food Freezing and Thawing Calculations,*
SpringerBriefs in Food, Health, and Nutrition, DOI 10.1007/978-1-4939-0557-7_7,
© The Author 2014

product microstructure and the freezing regime. The effect of thermal conductivity is more pronounced at large Biot numbers (see Sect. 3.2.3).

For quick calculations, engineers still have to rely on an approximate or empirical method. Of those reviewed in Chap. 4, Pham's (1984) Method 1 is the most reliable, followed by Pham's (1986) Method 2, although the correction factor given in Sect. 4.2.8 should be applied for low moisture, low freezing point foods and/or when cryogenic temperatures are used.

Hardly any freezing problem is "thermal only", and more attention will be devoted in the future to solving for the effect of parallel coupled physical processes: mass transfer, nucleation, crystal growth, mass transfer across cell membranes, vitrification, thermal expansion, mechanical strain and stress, cracking. Even in normal freezing, internal pressure may have some effect on the freezing point that has been neglected up till now. The modern food engineer is no longer interested only in freezing times or heat loads, but also in food quality factors: drip, color, texture, flavor, distortion and cracks and microbial growth (especially during thawing). To predict these factors, detailed modelling is needed on physical processes other than heat transfer.

While conventional FDM, FEM and FVM can deal with any continuous deterministic phenomenon that can be described by PDEs, some phenomena such as nucleation, crystal growth, crack initiation and crack growth are by nature discrete and stochastic, and thus the PDE approach may need to be augmented by another modelling approach altogether, such as Monte Carlo, cellular automata or hybrid automata models.

To model non-thermal phenomena successfully, data on some food properties hitherto neglected by food technologists (moisture diffusivity, absorption isotherm, nucleation parameters, cell size, cell membrane permeability, viscoelastic properties, tortuosity factor in porous foods, etc.) will have to be collected. More and better data on heat transfer coefficients and thermal properties and better methods for their prediction will also be an ongoing area of research. The use of computational fluid dynamics (CFD) for calculating heat transfer coefficients in food refrigeration is increasingly popular, but the lack of a satisfactory turbulence model for many practical situations (circulating flows, natural or mixed convection) means that CFD results cannot yet be completely trusted.

Nowadays, computers are so fast that it could be believed that there is no need to search for more efficient methods. However, the numerical modelling of the freezing of 3-D objects still takes many minutes or hours even when secondary effects such as crystal growth of mass transfer are ignored. Freezing models would also be more useful if they could be incorporated in larger programs such as models of whole food plants. Furthermore, the food engineer does not model for the fun of it but with the ultimate objective of being able to optimize products and processes. Computer optimization involves running the model hundreds or thousands of times (in the case of stochastic optimization methods such as genetic algorithms, even tens or hundreds of thousands). Models are also used in the determination of product properties and other parameters by error minimization, where they have to be run a similar number of times. Therefore, the search for more efficient algorithms will continue, even in this day and age of fast computers.

During food freezing, physical processes happen on widely differing scales, all of which may have important consequences: on the macro scale heat flow takes place through the whole product, on the meso scale (near the surface) moisture is exchanged between the food and the surrounding, on the micro scale heat, moisture and solutes are exchanged between intra- and extra-cellular spaces, and on the molecular scale nucleation and crystal growth occur and cause mechanical damage or trap solutes during freeze concentration. There is a need to develop multiscale models that would take all relevant phenomena into account and be able to predict accurately the effect of freezing processes on food quality.

References

Adalsteinsson, D., Sethian, J.A. (1999) The fast construction of extension velocities in level set methods. *Journal of Computational Physics*, **148**, 2–22.

AgResearch (2013) Food Product Modeller. www.mirinz.org.nz/prod/foodprodmod.asp, accessed 14 Dec 2013.

Añon, M.C., Calvelo, A. (1980) Freezing rate effects on the drip losses of frozen beef. *Meat Science* **4**, 1–14.

ANSYS (2005) ANSYS Thermal Analysis Guide, Ansys Release 10.0.

Arroyo, J.G., Mascheroni, R.H. (1990) A generalized method for the prediction of freezing times of regular and irregular foods. *Progress in the Science and Technology of Refrigeration in Food Engineering*, Proceedings of Meetings of Commissions B2, C2, D1, D2-D3, September 24–28, 1990, Dresden, pp. 643–649. Pub. International Institute of Refrigeration, Paris.

Auleda, J.M., Raventós, M., Hernández, E. (2011) Calculation method for designing a multi-plate freeze-concentrator for concentration of fruit juices. *Journal of Food Engineering*, **107**, 27–35.

Avrami, M. (1939) Kinetics of phase change I. *Journal of Chemical Physics*, **7**, 1103–1112.

Avrami, M. (1940) Kinetics of phase change II. *Journal of Chemical Physics*, **8**, 212–224.

Avrami, M. (1941) Kinetics of phase change III. *Journal of Chemical Physics*, **9**, 177–184.

Ayappa, K.G., Davis, H.T., Crapiste, G., Davis, E.A., Gordon, J. (1991) Microwave heating: an evaluation of power formulations. Chemical Engineering Science, **46**, 1005–1016.

Ayappa, K.G., Davis, H.T., Barringer, S.A., Davis, E.A. (1997) Resonant microwave power absorption in slabs and cylinders. *AIChE Journal*, **43**, 615–624.

Banaszek, J. (1989) Comparison of control volume and Galerkin finite element methods for diffusion-type problems. *Numerical Heat Transfer*, 59–78.

Bart G.C.J., Hanjalic K. (2003) Estimation of shape factor for transient conduction. *International Journal of Refrigeration*, **26**, 360–367.

Bayindirli, L., Ozilgen, M., Ungan, S. (1993) Mathematical analysis of freeze concentration of apple juice. *Journal of Food Engineering*, **19**, 95–107.

Becker, B.R., Fricke, B.A. (2004) Heat transfer coefficients for forced-air cooling and freezing of selected foods. *International Journal of Refrigeration*, **27**, 540–551.

Bennon, W.D., Incropera, F.P. (1987) A continuum model for momentum, heat and species transport in binary solid-liquid phase change systems—I. Model formulation. *International Journal of Heat and Mass Transfer*, **30**, 2161–2170.

Bevilacqua, A., Zaritsky, N.E., Calvelo, A. (1979) Histological mesurements of ice in frozen beef. *Journal of Food Technology* **1**, 237–251.

Bird R.B., Stewart W.E., Lightfoot E.N. (1960) *Transport Phenomena*. Wiley, New York.

Boettinger, W.J., Warren, J.A., Beckermann, C., Karma, A. (2002) Phase-field simulation of solidification. *Annual Review of Materials Research*, **32**, 163–194.

Bonacina, C., Comini, G. (1971) On a numerical method for the solution of the unsteady-state heat conduction equation with temperature-dependent parameters. *Proceedings 13th International Congress of Refrigeration*, Washington D.C., USA, Vol. 2, p. 329.

Botte G.G., Ritter J.A., White R.E. (2000) Comparison of finite difference and control volume methods for solving differential equations. *Computers and Chemical Engineering* **24**, 2633–2654.

Brown, S.G.R. (1998) Simulation of diffusional composite growth using the cellular automaton finite difference (CAFD) method. *Journal of Materials Science* **33**, 4769–4773.

Browne, D.J., Hunt, J.D. (2004) A fixed grid front-tracking model of the growth of a columnar front and an equiaxed grain during solidification of an alloy. *Numerical Heat Transfer, Part B: Fundamentals*, **45**, 395–419.

Budd, C.J., Hill, A.D.C. (2011) A comparison of models and methods for simulating the microwave heating of moist foodstuffs. *International Journal of Heat and Mass Transfer*, **54**, 807–817.

Budhia, H., Kreith, F. (1973) Heat transfer with melting or freezing in a wedge. *International Journal of Heat and Mass Transfer*, **16**, 195–211.

Caginalp, G. (1986) An analysis of a phase field model of a free boundary. *Archive for Rational Mechanics and Analysis*, **92**, 205–245.

Caginalp, G. (1989) Stefan and Hele-Shaw type models as asymptotic limits of the phase-field equations. *Physical Review A*, **39**, 5887–5896.

Caginalp, G., Chen, X. (1998) Convergence of the phase field model to its sharp interface limits. *European Journal of Applied Mathematics*, **9**, 417–445.

Califano, A.N., Zaritzky, N.E. (1997) Simulation of freezing or thawing heat conduction in irregular two dimensional domains by a boundary fitted grid method. *Lebensmittel-Wissenschaft und-Technologie* **30**, 70–76.

Campañone, L.A., Salvadori, V.O., Mascheroni, R.H. (2001) Weight loss during freezing and storage of unpackaged foods. *Journal of Food Engineering* **47**, 69–79.

Carslaw, H.S., Jaeger, J.C. (1959) *Conduction of Heat in Solids*, 2nd edn, Clarendon Press, Oxford.

CASC (2013), Mathematical Software, Centre for Applied Computing, Lawrence Livermore National Laboratory, https://computation.llnl.gov/casc/software.html. Accessed 12 Sep 2013.

Chang, A., Dantzig, J.A., Darr, B.T., Hubel, A. (2007) Modeling the interaction of biological cells with a solidifying interface. *Journal of Computational Physics* **226**, 1808–1829.

Chen, P., Chen, X.D. (2000) A generalized correlation of solute inclusion in ice formed from aqueous solutions and food liquids on sub-cooled surface. *The Canadian Journal of Chemical Engineering*, **78**, 312–319.

Chen, X.D., Chen, P., Free, K.W. (1997) A note on the two models of ice growth velocity in aqueous solutions derived from an irreversible thermodynamics analysis and the conventional heat and mass transfer theory. *Journal of food engineering*, **31**, 395–402.

Chen, S., Merriman, B., Osher, S., Smereka, P. (1997) A simple level set method for solving Stefan problems. *Journal of Computational Physics*, **135**, 8–29.

Choi, Y., Okos, M.R. (1986) Effects of temperature and composition on the thermal properties of foods. In *Food Engineering and Process Applications*, Vol. 1: *Transport Phenomena*, pp. 93–101. Elsevier, New York.

Choudhury, A., Reuther, K., Wesner, E., August, A., Nestler, B., Rettenmayr, M. (2012) Comparison of phase-field and cellular automaton models for dendritic solidification in Al-Cu alloy. *Computational Materials Science*, **55**, 263–268.

Chourot, J.M., Boillereaux, L., Havet, M., Le Bail, A. (1997) Numerical modeling of high pressure thawing: Application to water thawing. *Journal of Food Engineering*, **34**, 63–75.

Cleland, A.C. (1977) Heat transfer during freezing of foods and prediction of freezing times. PhD thesis, Massey University, New Zealand.

Cleland, A.C. (1985) RADS—a computer package for refrigeration analysis, design and simulation. *International Journal of Refrigeration*, **8**, 372–373.

Cleland, A.C. (1990) *Food Refrigeration Processes Analysis, Design and Simulation*. Elsevier, London.

Cleland, A.C., Earle, R.L. (1976) A comparison of freezing calculations including modification to take into account initial superheat. *Refrigeration Science and Technology* (International Institute of Refrigeration, Meeting Commissions C2, D1, D2, D3 and E1, Melbourne 6–10 Sep), **1976(1)**, 369–376.

Cleland, A.C., Earle, R.L. (1977) A comparison of analytical and numerical methods for predicting the freezing times of foods. *Journal of Food Science*, **42**, 1390–1395.

Cleland, A.C., Earle, R.L. (1979a) A comparison of methods for predicting the freezing times of cylindrical and spherical foodstuffs. *Journal of Food Science*, **44**, 958–963.

Cleland, A.C., Earle, R.L. (1979b) Prediction of freezing times for foods in rectangular packages. *Journal of Food Science*, **44**, 964–970.

Cleland, A.C., Earle, R.L. (1982) A simplified method for prediction of heating and cooling rate in solids of various shapes. *International Journal of Refrigeration*, **5**, 98–106.

Cleland, A.C., Earle, R.L. (1984a) Freezing time prediction for different final product temperatures. *Journal of Food Science*, **49**, 1230–1232.

Cleland, A.C., Earle, R.L. (1984b) Assessment of freezing time prediction methods. *Journal of Food Science*, **49**, 1034–1042.

Cleland, A.C., Earle, R.L., Cleland, D.J. (1982) The effect of freezing rate on the accuracy of numerical freezing calculations. *International Journal of Refrigeration*, **5**:294–301.

Cleland, D.J. (1991) A generally applicable simple method for prediction of food freezing and thawing times. Proceedings 18th International Congress of Refrigeration, Montreal, Elsevier, London, Vol. **4**, pp. 1873–1877.

Cleland, D.J., Cleland, A.C., Earle, R.W., Byrne, S.J. (1984) Prediction of rates of freezing, thawing and cooling in solids of arbitrary shape using the finite element method. *International Journal of Refrigeration*, **7**, 6–13.

Cleland, D.J., Cleland, A.C., Earle, R.L., Byrne, S.J. (1986) Prediction of thawing time for foods of simple shape. *International Journal of Refrigeration*, **9**, 220–228.

Cleland, D.J., Cleland, A.C., Earle, R.L. (1987a) Prediction of freezing & thawing times for multidimensional shapes by simple formulae. Part 1: Regular shapes. *International Journal of Refrigeration*, **10**, 157–164.

Cleland, D.J., Cleland, A.C., Earle, R.L. (1987b) Prediction of freezing & thawing times for multidimensional shapes by simple formulae. Part 2: Irregular shapes. *International Journal of Refrigeration*, **10**, 234–240.

Comini, G., Del Giudice, S. (1976) Thermal aspects of cryosurgery. *Journal of Heat Transfer*, **98**, 543–549.

Comini, G., Del Giudice, S., Saro, O. (1989) Conservative equivalent heat capacity methods for non-linear heat conduction, in *Numerical Methods in Thermal Problems* (eds R.W. Lewis and K. Morgan), Pineridge Press, Swansea, Vol. 6, Part 1, pp. 5–15.

Crank, J., Nicolson, P. (1947) A practical method for numerical integration of solutions of partial Nicolson differential equations of heat conduction type. *Proceedings of Cambridge Philosophical Society*, **43**, 50–67.

Davey, L.M., Pham, Q.T. (1997) Predicting the dynamic product heat load and weight loss during beef chilling using a multi-region finite difference approach. *International Journal of Refrigeration*, **20**, 470–482.

de Michelis, A., Calvelo, A. (1982) Mathematical models for non-symmetric freezing of beef. *Journal of Food Science*, **47**, 1211–1217.

de Michelis, A., Calvelo, A. (1983) Freezing time prediction methods for bricks and cylindrical shaped foods. *Journal of Food Science*, **48**, 909–913.

De Vries, D.A. (1952) The thermal conductivity of granular materials, *Annex 1952-1 Bulletin of the International Institute of Refrigeration*, pp. 115–131.

De Vries, D.A. (1958) Simultaneous transfer of heat and moisture in porous media. *Transactions of the American Geophysical Union*, **39**, 909–916.

De Vries, D.A. (1987) The theory of heat and moisture transfer in porous media revisited. *International Journal of Heat and Mass Transfer*, **30**, 1343–1350.

Denys, S., Van Loey, A.M., Hendrickx, M.E. (1997) Modeling heat transfer during high-pressure freezing and thawing. *Biotechnology Progress*, **13**, 416–423.

Denys, S., Van Loey, A.M., Hendrickx, M.E. (2000) Modeling conductive heat transfer during high-pressure thawing processes: Determination of latent heat as a function of pressure. *Biotechnology Progress*, **16**, 447–455.

Devireddy, R.V., Smith, D.J., Bischof, J.C. (2002) Effect of microscale mass transport and phase change on numerical prediction of freezing in biological tissues. *Journal of Heat Transfer*, **124**, 365–374.

Douglas J. Jr. (1962) Alternating direction methods for three space variables. *Numerische Mathematik* **4**(1): 41–63.

Duckworth, R.B. (1971) Differential thermal analysis of frozen food systems: the determination of unfreezable water. *International Journal of Food Science & Technology*, **6**, 317–327.

Eyres, N.R., Hartree, D.R., Ingham, J., Jackson, R., Sarjant, R.J., Wagstaff, J.B. (1946) The calculation of variable heat flow in solids. *Transactions of the Royal Society*, **A240**:1.

Fabbri, M., Voller, V.R. (1997) The Phase-Field Method in the Sharp-Interface Limit: A Comparison between Model Potentials. *Journal of Computational Physics*, **130**, 256–265.

Fenman, R.P. (2013) *The Feynman Lectures on Physics*, Chapter 18: The Maxwell Equations. http://feynmanlectures.caltech.edu/II_18.html, accessed 3 January, 2014.

Fikiin, K.A. (1996) Generalised numerical modelling of unsteady heat transfer during cooling and freezing using an improved enthalpy method and quasi-one-dimensional formulation. *International Journal of Refrigeration*, **19**, 132–140.

Fikiin, K.A. (1998) Ice content prediction methods during food freezing: a survey of the Eastern European literature. *Journal of Food Engineering*, **38**, 331–339.

Fikiin, K.A., Fikiin, A.G. (1998) Individual quick freezing of foods by hydrofluidisation and pumpable ice slurries, In *Advances in the Refrigeration Systems, Food Technologies and Cold Chain, Proceedings of IIR Conference, Sofia (Bulgaria), Refrigeration Science and Technology, International Institute of Refrigeration*, pp. 319–326. Republished in *AIRAH Journal* (2001) **55**(11), 15–18.

Fikiin, K.A., Fikiin, A.G. (1999) Predictive equations for thermophysical properties and enthalpy during cooling and freezing of food materials. *Journal of Food Engineering*, **40**, 1–6.

Fikiin, K., Tsvetkov, O., Laptev, Yu., Fikiin, A., Kolodyaznaya, V (2003) Thermophysical and engineering issues of the immersion freezing of fruits in ice slurries based on sugar-ethanol aqueous solution. *Ecolibrium* August, 10–14.

Flesland, O. (1995) Freeze concentration by layer crystallization. *Drying Technology*, **13**, 1713–1739.

Franks, F. (2003) Nucleation of ice and its management in ecosystems. *Philosophical Transactions of the Royal Society London A*, **361**, 557–574

Fricke B.A., Becker B.R. (2001) Evaluation of thermophysical property models for foods. *HVAC&R Research*, **7**, 311–329.

Gibou, F., Fedkiw, R., Caflisch, R., Osher, S. (2003) A Level Set Approach for the Numerical Simulation of Dendritic Growth. *Journal of Scientific Computing*, **19**, 183–199.

Gold, L.W. (1958) Some observations on the dependence of strain on stress in ice. *Canadian Journal of Physics*, **36**, 1265–1275.

Gu, A., Suzuki, T., Miyawaki, O. (2005) Limiting Partition Coefficient in Progressive Freeze-concentration. *Journal of Food Science*, **70**, 546–551.

Gueyffier, D., Li, J., Nadim, A., Scardovelli, R., Zaleski, S. (1999) Volume-of-fluid interface tracking with smoothed surface stress methods for three-dimensional flows. *Journal of Computational Physics*, **152**, 423–456.

Gulati, T., Datta, A.K. (2013) Enabling computer-aided food process engineering: Property estimation equations for transport phenomena-based models. *Journal of Food Engineering*, **116**, 483–504.

Hamdami, N., Monteau, J.Y., Le Bail, A. (2003) Effective thermal conductivity of a high porosity model food at above and sub-freezing temperatures. *International Journal of Refrigeration*, **26**, 809–816.

Hamdami, N., Monteau, J.Y., Le Bail, A. (2004a) Heat and mass transfer simulation during freezing in bread, in *Proceedings ICEF9 9th International Congress on Engineering and Foods, Montpellier* (CDROM).

Hamdami, N., Monteau, J.Y., Le Bail, A. (2004b) Simulation of coupled heat and mass transfer during freezing of a porous humid matrix. *International Journal of Refrigeration*, **27**, 595–603.

Hamdami, N., Monteau, J.Y., Le Bail, A. (2004c) Heat and mass transfer in par-baked bread during freezing. *Food Research International*, 37, 477–488.

Hamdami, N., Pham, Q.T., Le Bail, A., Monteau, J.Y. (2007) Two-stage freezing of part baked breads: application and optimization. *Journal of Food Engineering*, 82, 418–426.

Hare, D.E., Sorensen, C.M. (1987) The density of supercooled water. II. Bulk samples cooled to the homogeneous nucleation limit. *Journal of Chemical Physics*, 87, 4840–4845.

Hayakawa, K., Villalobos, G. (1989) Formulas for estimating Smith *et al.* parameters to determine mass average temperature of irregular shaped bodies. *Journal of Food Process Engineering*, 11, 237–256.

Heldman, D.R., Gorby, D.P. (1975) Prediction of thermal conductivity in frozen food. *Transactions ASAE* 18:740–744.

Holten, V., Bertrand, C.E., Anisimov, M.A., Sengers, J.V. (2013) Thermodynamics of supercooled water. *Journal of Chemical Physics*, 136, 094507.

Horvay, G., Cahn, J.W. (1961) Dendritic and spherical growth. Acta Metallurgica 9, 695–705.

Hossain, Md.M., Cleland, D.J., Cleland, A.C. (1992a) Prediction of freezing & thawing times for foods of regular multidimensional shape by using an analytically derived geometric factor. *International Journal of Refrigeration*, 15, 227–234.

Hossain, Md.M., Cleland, D.J., Cleland, A.C. (1992b) Prediction of freezing and thawing times for foods of two-dimensional irregular shape by using a semi-analytical geometric factor. *International Journal of Refrigeration*, 15, 235–240.

Hossain, Md.M., Cleland, D.J., Cleland, A.C. (1992c) Prediction of freezing and thawing times for foods of three-dimensional irregular shape by using a semi-analytical geometric factor. *International Journal of Refrigeration*, 15, 241–246.

Hu, Z., Sun, D-W (2001) Predicting local surface heat transfer coefficients by different turbulent k-ϵ models to simulate heat and moisture transfer during air-blast chilling. *International Journal of Refrigeration*, 24, 702–717.

Huige, N.J.J., Thijssen, H.A.C. (1972) Production of large crystals by continuous ripening in a stirred tank. *Journal of Crystal Growth*, 13, 483–487.

Hung, Y.C., Thomson, D.R. (1983) Freezing time prediction for slab shaped foodstuffs by an improved analytical method. *Journal of Food Science*, 48, 555–560.

Ilicali C, Holacar M (1990) A simplified approach for predicting the freezing times of foodstuffs of anomalous shape, in *Engineering and Food*, Vol. 2, (eds W.E.L. Spiess and H. Schubert), Elsevier, London, pp. 418–425.

Ilicali C, Saglam N, (1987) A simplified analytical model for freezing time calculation in foods. *Journal of Food Process Engineering* 9, 299.

Ilicali C, Tang HT, Lim PS (1999) Improved Formulations of Shape Factors for the Freezing and Thawing Time Prediction of Foods. *LWT—Food Science and Technology*, 32, 312–315.

Irimia, D., Karlsson, J.O.M (2002) Kinetics and mechanism of intercellular ice propagation in a micropatterned tissue construct. *Biophysical Journal*, 82, 1858–1868.

Irimia, D., Karlsson, J.O.M (2005) Kinetics of Intracellular Ice Formation in One-Dimensional Arrays of Interacting Biological Cells. *Biophysical Journal*, 88, 647–660.

Ivantsov, G.P. (1947) Temperature field around spherical, cylindrical and needle-shaped crystals which grow in supercooled melt. *Doklady Akademii Nauk SSSR* 58, 567–569.

Jarvis, D.J., Brown, S.G.R., Spittle, J.A. (2000) Modelling of non-equilibrium solidification in ternary alloys: comparison of 1D, 2D and 3D cellular automaton-finite difference simulation. *Materials Science and Technology*, 16, 1420–1424.

Jaeger, M., Carin, M. (2002) The front-tracking ALE method: application to a model of the freezing of cell suspensions. *Journal of Computational Physics*, 179, 704–735.

Jaeger, M., Carin, M., Medale, M., Tryggvason, G. (1999) The osmotic migration of cells in a solute gradient. *Biophysics Journal* 77, 1257–1267.

Karagadde, S., Bhattacharya, A., Tomar, G., Dutta, P. (2012) A coupled VOF-IBM-enthalpy approach for modeling motion and growth of equiaxed dendrites in a solidifying melt. *Journal of Computational Physics*, 231, 3987–4000.

Karel, M., Lund, D.B. (2003) *Physical Principles of Food Preservation*, 2nd edn. Marcel Dekker, New York.

Kell, G.S. (1975) Density, thermal expansivity, and compressibility of liquid water from 0 to 150°C. *Journal of Chemical and Engineering Data*, **20**, 97–105.

Kirchhoff, G. (1894) *Vorlesungen uber die Theorie der Warme*, quoted in H.S. Carslaw and J.C. Jaeger (1986) *Conduction of Heat in Solids*, 2nd edn, Clarendon Press, Oxford, p. 11.

Kobayashi, R. (1993) Modeling and numerical simulations of dendritic crystal growth. *Physica D*, **63**, 410–423.

Kondjoyan, A. (2006) A review on surface heat and mass transfer coefficients during air chilling and storage of food products. *International Journal of Refrigeration*, **29**, 863–875.

Lacroix, C, Castaigne, F (1988) Freezing time calculation for products with simple geometrical shapes. *Journal of Food Process Engineering* **10**, 81–104.

Le Bail, A., Chevalier, D., Mussa, D.M., Ghoul, M. (2002) High pressure freezing and thawing of foods: a review. *International Journal of Refrigeration*, **25**, 504–513.

Le Bail, A., Monteau, J.-Y., Margerie, F., Lucas, T., Chargelegue, A., Reverdy, Y. (2005) Impact of selected process parameters on crust flaking of frozen part baked bread. *Journal of Food Engineering*, **69**, 503–509.

Lees, M. (1976) A linear three-level difference scheme for quasilinear parabolic equations, *Mathematics of Computation*, **20**, 516–522.

Lemmon, E.C. (1979) Phase change technique for finite element conduction code, in *Numerical Methods in Thermal Problems* (eds R.W. Lewis and K. Morgan), Pineridge Press, Swansea, pp. 149–158.

Levy, L. (1981) A modified Maxwell-Eucken equation for calculating the thermal conductivity of two-component solutions or mixtures. *International Journal of Refrigeration*, **4**, 223–225.

Lian, G., Moore, S., Heeney, L. (2006) Population balance and computational fluid dynamics modelling of ice crystallisation in a scraped-surface freezer. *Chemical Engineering Science*, **61**, 7819–26.

Lide, D.R. (2009) *CRC Handbook of Chemistry and Physics*, 90th Edn. CRC Press, Boca Raton.

Lin, Z., Cleland, A.C., Sellarach, G.F., Cleland, D.J. (1993) Prediction of chilling times for objects of regular multi-dimensional shapes using a general geometric factor. In Refrigeration Science and Technology 1993-3. International Institute of Refrigeration, Paris, pp. 259–267.

Lin, Z., Cleland, A.C., Sellarach, G.F., Cleland, D.J. (1996a) A simple method for prediction of chilling times for objects of two-dimensional irregular shape. *International Journal of Refrigeration*, **19**, 95–106.

Lin, Z., Cleland, A.C., Sellarach, G.F., Cleland, D.J. (1996b) A simple method for prediction of chilling times: Extension to three-dimensional irregular shapes. *International Journal of Refrigeration*, **19**, 107–114. Erratum (2000), *International Journal of Refrigeration*, **23**, 168.

Lovatt, S.J., Pham, Q.T., Loeffen, M.P.F., Cleland, A.C. (1993a) A new method of predicting the time-variability of product heat load during food cooling, Part 1: Theoretical considerations. *Journal or Food Engineering*, **18**, 13–36.

Lovatt, S.J., Pham, Q.T., Loeffen, M.P.F., Cleland, A.C. (1993b) A new method of predicting the time-variability of product heat load during food cooling, Part 2: Experimental testing. *Journal or Food Engineering*, **18**, 37–62.

Lovatt, S.J., Loeffen, M.P.F., Cleland, A.C. (1998) Improved dynamic simulation of multi-temperature industrial refrigeration systems for food chilling, freezing and cold storage. *International Journal of Refrigeration*, **21**, 247–260.

Lucas, T., Flick, D., Raoult-Wack, A.L. (1999) Mass and thermal behaviour of the food surface during immersion freezing. *Journal of Food Engineering*, **41**, 23–32.

Lucas, T., Chourot, J.M., Bohuon, Ph., Flick, D. (2001) Freezing of a porous medium in contact with a concentrated aqueous freezant: numerical modelling of coupled heat and mass transport. *International Journal of Heat and Mass Transfer*, **44**, 2093–2106.

Mao, L., Udaykumar, H.S., Karlsson, J.O.M. (2003) Simulation of micro-scale interaction between ice and biological cells. *International Journal of Heat and Mass Transfer*, **46**, 5123–5136.

Martino, M.N., Zaritzky, N.E. (1988) Ice crystal size modification during frozen beef storage. *Journal of Food Science* **53**, 1631–1637.

Martino, M.N., Zaritzky, N.E. (1989) Ice recrystallization in a model system and in frozen muscle tissue. *Cryobiology* **26**, 138–148.

Mascheroni, R.H., Calvelo, A. (1982) A simplified model for freezing time calculations in foods. *Journal of Food Science*, **47**, 1201–1207.

Mascheroni, R.H., Ottino, J., Calvelo, A. (1977) A model for the thermal conductivity of frozen meat. *Meat Science*, **1**, 235–243.

Mazur, P. (1963) Kinetics of water loss from cells at subzero temperature and the likelihood of intracellular freezing. *Journal of General Physiology*, **47**, 347–69.

McKellar, A., Paterson, J., Pham, Q.T. (2009) A Comparison of Two Models for Stresses and Strains during Food Freezing. *Journal of Food Engineering*, **95**, 142–150.

McNabb, A., Wake, G.C., Hossain, Md.M., Lambourne, R.D. (1990a) Transition times between steady states for heat conduction, Part I: General theory and some exact results. *Occasional Pubs in Maths & Statistics* No.20, Massey University.

McNabb, A., Wake, G.C., Hossain, Md.M., Lambourne, R.D. (1990b) Transition times between steady states for heat conduction, Part II: Approximate solutions and examples. *Occasional Pubs in Maths & Statistics* No.21, Massey University.

Menegalli, F.C., Calvelo, A. (1978) Dendritic growth of ice crystals during freezing of beef. *Meat Science*, **3**,179–199.

Michelmore, R. W., Franks, F. (1982) Nucleation rates of ice in undercooled water and aqueous solutions of polyethylene glycol. *Cryobiology*, **19**, 163–171.

Michon, G.P. (2013) Final Answers, http://www.numericana.com/answer/ellipsoid.htm, accessed 18-Jan-2013.

Miles, C.A., Mayer, Z., Morley, M.J., Houska, M. (1997) Estimating the initial freeezing point of foods from composition data. *International Journal of Food Science and Technology*, **32**, 389–400.

Miyawaki, O. (2001) Progressive freeze-concentration: a new method for high quality concentration of liquid food. *Food Engineering Progress*, **5**, 190–194.

Miyawaki, O., Abe, T., Yano, T. (1989) A numerical model to describe freezing of foods when supercooling occurs. *Journal of Food Engineering*, **9**, 143–151.

Miyawaki, O., Liu, L., Shirai, Y., Sakashita, S., Kagitani, K. (2005) Tubular ice system for scale-up of progressive freeze-concentration. *Journal of Food Engineering*, **69**, 107–113.

Moelans, N., Blanpain, B., Wollants, P. (2008) An introduction to phase-field modeling of microstructure evolution. *Computer Coupling of Phase Diagrams and Thermochemistry*, **32**, 268–294.

Morgan, K., Lewis, R.W., Zienkiewicz, O.C. (1978) An improved algorithm for heat conduction problems with phase change. *International Journal of Numererical Methods in Engineering*, **12**, 1191–1195.

Müller, M., Sekoulov, I. (1992) Waste water reuse by freeze concentration with a falling film reactor. *Water Science and Technology*, **26**, 1475–1482.

Müller, G., Métois, J., Rudolph, P. (2004) *Crystal growth from fundamentals to technology*, Elsevier, Amsterdam, The Netherlands.

Nahid, A, Bronlund JE, Cleland DJ, Oldfield DJ, Philpott B. (2004) Prediction of thawing and freezing of bulk palletised butter, *Proceedings of the 9th International Congress on Engineering and Foods (ICEF 9), Montpellier* (CDROM), paper 668.

Nahid, A., Bronlund, J.E., Cleland, D.J., Philpott, B.A. (2006) Modelling the freezing of single cartons of butter, *Proceedings 11th Asian Pacific Confederation of Chemical Engineering (APCChe) Congress*, Kuala Lumpur, Malaysia, 27–30/8/2006.

Nahid, A., Bronlund, J.E., Cleland, D.J., Philpott, B.A. (2008) Modelling the freezing of butter. *International Journal of Refrigeration*, **31**, 152–160.

Netlib (2013) ODE Solver Repository, http://www.netlib.org/ode/index.html. Accessed 12 Sep 2013.

NIST (2013a) Thermophysical Properties of Fluid Systems.//webbook.nist.gov/chemistry/fluid/, accessed 22 Dec 2013.

NIST (2013b) GAMS Guide to Available Mathematical Software.//gams.nist.gov/. Accessed 12 Sep 2013.

Norton, T., Delgado, A., Hogan, E., Grace, P. & Sun, D.-W. (2009) Simulation of high pressure freezing processes by enthalpy method. *Journal of Food Engineering*, **91**, 260–268.

Norton, T., Sun, D.-W. (2013) An overview of CFD applications in the food industry, in *Computational Fluid Dynamics in Food Processing*, ed. D.-W. Sun, CRC Press, New York, pp. 2–41.

Oliveira, M.E.C., Franca, A.S. (2002) Microwave heating of foodstuffs. *Journal of Food Engineering*, **53**, 347–359.

Osher, S.,, Fedkiw, R. (2001) Level set methods: An overview and some recent results. *Journal of Computational Physics*, **169**, 475–502.

Osher, S., Sethian, J.A. (1988) Front propagating with curvature-dependent speed: Algorithms based on Hamilton-Jacobi formulation. *Journal of Computational Physics*, **79**, 12–49.

Otero, L., Sanz, P. (2000) High pressure shift freezing, Part 1: Amount of ice instantaneously formed in the process. *Biotechnology Progress*, **16**, 1030–1036.

Otero, L., Ousegui, A., Guignon, B., Le Bail, A., Sanza, P.D. (2006) Evaluation of the thermophysical properties of Tylose gel under pressure in the phase change domain. *Food Hydrocolloids* **20**, 449–460.

Pal, D., Bhattacharya, J., Dutta, P. & Chakraborty, S. (2006): An enthalpy model for simulation of dendritic growth. *Numerical Heat Transfer, Part B: Fundamentals*, **50**, 59–78.

Patankar, S. (1980). *Numerical heat transfer and fluid flow*. Hemisphere Publishing, New York.

Peaceman, D.W., Rachford, H.H. (1955) The numerical solution of parabolic and elliptic differential equations. *Journal of the Society for Industrial and Applied Mathematics*, **3**, 28–41.

Pedroso, R.I., Domoto, A. (1973) Inward spherical solidification-solution by the method of strained coordinates. *International Journal of Heat and Mass Transfer*, **16**, 1037–1043.

Petrenko, V.F., Whitworth, R.W. (1999) *Physics of Ice*. Oxford University Press, Oxford, UK.

Pflug, I.J., Blaisdell, J.L., Kopelman, I.J. (1965) Developing temperature-time curves for objects that can be approximated by a sphere, infinite plate or infinite cylinder. *ASHRAE Transactions*, **71**, 238–248.

Pham, Q.T. (1984) An extension to Plank's equation for predicting freezing times of foodstuffs of simple shapes. *International Journal of Refrigeration*, **7**, 377–383.

Pham QT (1985a) An analytical method for predicting freezing times of rectangular blocks of foodstuffs. *International Journal of Refrigeration*, **8**, 43–47.

Pham, Q.T. (1985b) A fast, unconditionally stable finite-difference method for heat conduction with phase change. *International Journal of Heat and Mass Transfer*, **28**, 2079–2084.

Pham Q.T. (1986a) Simplified equation for predicting the freezing time of foodstuffs. *Journal of Food Technology*, **21**, 209–221.

Pham, Q.T. (1986b) Freezing of foodstuffs with variations in environmental conditions. *International Journal of Refrigeration*, **9**, 290–295.

Pham, Q.T. (1986c) The use of lumped capacitances in the finite-element solution of heat conduction with phase change. *International Journal of Heat and Mass Transfer*, **29**, 285–292.

Pham, Q.T. (1987a) Calculation of bound water in frozen foods. *Journal of Food Science* **52**, 210–212.

Pham, Q.T. (1987b) A converging-front model for the asymmetric freezing of slab-shaped foodstuffs. *Journal of Food Science*, **52**, 795–800.

Pham, Q.T. (1987c) A comparison of some finite-difference methods for phase-change problems, *Proceedings 17th International Congress of Refrigeration*, Vienna, vol. B, pp. 267–271.

Pham, Q.T. (1989) Effect of supercooling on freezing times due to dendritic growth of ice crystals. *International Journal of Refrigeration*, **12**, 295–300.

Pham, Q.T. (1991) Shape factor for the freezing time of ellipses and ellipsoids. *Journal of Food Engineering*, **13**, 159–170.

Pham, Q.T. (1995) Comparison of general purpose finite element methods for the Stefan problem. *Numerical Heat Transfer Part B—Fundamentals* 27:417–435.

Pham, Q.T. (1996) Prediction of calorimetric properties and freezing time of foods from composition data. *Journal of Food Engineering*, **30**, 95–107.

Pham, Q.T. (2001) Prediction of cooling/freezing/thawing time and heat load, in: *Advances in Food Refrigeration* (ed. D.-W. Sun) Leatherhead Food RA, Leatherhead, pp. 110–152.

Pham, Q.T. (2006) Modelling heat and mass transfer in frozen foods: a review. *International Journal of Refrigeration*, **29**, 876–888.

Pham, Q.T. (2014) Freezing time formulas for foods with low moisture content, low freezing point and for cryogenic freezing. *Journal of Food Engineering*, **127**, 85–92.

Pham, Q.T., Karuri, N. (1999) A computationally efficient technique for calculating simultaneous heat and mass transfer during food chilling, *20th International Congress of Refrigeration*, Sydney, 19–24 Sep 1999.

Pham, Q.T., Mawson, R. (1997) Moisture migration and ice recrystallization in frozen foods, in *Quality in Frozen Foods* (ed. M.C. Erickson and Y.C. Hung), Chapman and Hall, New York, pp. 67–91.

Pham Q.T., Willix, J. (1984) A model for food desiccation in frozen storage. *Journal of Food Science*, **49**,1275–81, 1294.

Pham Q.T., Willix J. (1989) Thermal conductivity of fresh lamb meat, fat and offals in the range −40 to +30 °C: Measurements and correlations. *Journal of Food Science* **54**, 508–515.

Pham, Q.T., Willix, J. (1990) Effect of Biot number and freezing rate on the accuracy of some food freezing time prediction methods. *Journal of Food Science*, **55**, 1429–34.

Pham, Q.T., Wee, H.K., Kemp, R.M., Lindsay, D. (1994) Determination of enthalpy values of foods by an adiabatic calorimeter. *Journal of Food Engineering*, **21**: 137–156.

Pham Q.T., Sutjiadi M.R., Sagara Y., Do G.-S. (2006a) Determination of Thermal Conductivity of Frozen Meat by Finite Element Modelling. *Proceedings FoodSim Conference* 2006, Naples.

Pham, Q.T., Le Bail, A., Hayert, M., Tremeac, B. (2006b) Stresses and cracking in freezing spherical foods: a numerical model. *Journal of Food Engineering*, **71**, 408–418.

Pham, Q.T., Le Bail, A., Tremeac, B. (2006c). Analysis of stresses during the freezing of spherical foods. *International Journal of Refrigeration*, **29**, 125–133.

Pham, Q.T., Trujillo, F.J., McPhail, N. (2009) Finite Element Model for Beef Chilling Using CFD-Generated Heat Transfer Coefficients. *International Journal of Refrigeration*, **32**, 102–113.

Plank, R. (1913a) Beitrage zur Berechnung und Bewertung der Gefriergeschwindigkeit von Lebensmitteln, *Zeitschrift fur die gesamte Kalte Industrie*, Reihe **3** Heft 10, 1–16.

Plank, R. (1913b) Die Gefrierdauer von Eisblocken. *Zeitschrift fur die gesamte Kalte Industrie*, **20**(6), 109–114.

Press WH, Flannery BP, Vetterling WT, Teukolky SA (1986) *Numerical Recipes in C*, 2nd Edition. Cambridge U.P., Cambridge, pp. 50–51.

Price, P.H., Slack, M.R. (1954) The effect of latent heat on numerical solutions of the heat flow equation. *British Journal of Applied Physics*, **5**(8) 285–287.

Provatas, N. & Elder, K. (2010) *Phase-Field Methods in Materials Science and Engineering*. Wiley, New York.

Provatas, N., Goldenfeld, N., Dantzig, J.A. (1999) Adaptive mesh refinement computation of solidification microstructures using dynamic data structures. *Journal of Computational Physics*, **148**, 265–290.

Raabe, D. (2001) Mesoscale Simulation of Recrystallization Textures and Microstructures. *Advanced Engineering Materials*, **3**, 745–752.

Raabe, D. (2002) Cellular automata in materials science with particular reference to recrystallization simulation. *Annual Review of Materials Research*, **32**, 53–76.

Rabin, Y., Steif, P.S. (1998) Thermal stresses in a freezing sphere and its application in cryobiology. *Transactions of the ASME*, **65**, 328–333.

Rabin, Y., Taylor, M.J., Wolmark, N. (1998) Thermal expansion measurement of frozen biological tissues at cryogenic temperatures. *ASME Journal of Biomechanical Engineering*, **120**, 259–266.

Ramaswamy, H.S., Lo, K.V., Tung, M.A. (1982) Simplified equations for transient temperatures in conductive foods with convective heat transfer at surface. *Journal of Food Science*, **47**, 2042–2047.

Rappaz, M. & Gandin, Ch.A. (1993) Probabilistic modelling of microstructure formation in solidification processes. *Acta Metallurgica et Materialia*, **41** 345–360.

Ratkje, S.K., Flesland, O. (1995) Modeling the freeze concentration process by irreversible thermodynamics. *Journal of Food Engineering*, **25**, 553–567.

Riedel, L. (1960) Eine Prufsubstanz fur gefrierversuche. *Kaltetechnik*, **12**:222–225.

Riley, D.S., Smith, F.T., Poots, G. (1974) The inward solidification of spheres and circular cyinders. *International Journal of Heat and Mass Transfer*, **17**, 1507–1516.

Rolph, W.R., Bathe, K.J. (1982) An efficient algorithm for analysis of nonlinear heat transfer with phase changes. *International Journal for Numerical Methods in Engineering*, **18**, 119–134.

Roose, J., Storrer, O. (1984) Modelization of phase changes by fictitious-heat flow. *International Journal for Numerical Methods in Engineering*, **20**, 217–225.

Rubinsky, B., Cravalho, E.G., Mikic, B. (1980) Thermal stresses in frozen organs. *Cryobiology*, **17**, 66–73.

Rutter, J.W., Chalmers, B. (1953). A prismatic structure formed during solidification metals. *Canadian Journal of Physics*, **31**, 15–39.

Salvadori, V.O. (1994). Transferencia de calor durante la congelación, el almacenamiento y la descongelación de alimentos. PhD Thesis, Universidad Nacional de La Plata, La Plata, Argentina.

Salvadori, VO, Mascheroni, RH (1991) Prediction of freezing and thawing time of foods by means of a simplified analytical method. *Journal of Food Engineering*, **13**, 67–78.

Salvadori, VO, Masheroni, RH (1996) Freezing of strawberry pulp in large containers: experimental determination and prediction of freezing times. *International Journal of Refrigeration*, **19**, 87–94.

Salvi, D., Boldor, D., Ortego, J., Aita, G.M., Sabliov, C.M. (2010) Numerical modeling of continuous flow microwave heating: a critical comparison of COMSOL and ANSYS. Journal of Microwave Power and Electromagnetic Energy, 44, 187–197.

Scheerlinck, N. (2000) Uncertainty Propagation Analysis of Coupled and Non-Linear Heat and Mass Transfer Models, PhD Thesis, Katholieke Universteit Leuven, pp. 180–181.

Scheerlinck, N., Verboven, P., Fikiin, K.A., de Baerdemaeker, J., Nicolai, B.M. (2001) Finite-element computation of unsteady phase change heat transfer during freezing or thawing of food using a combined enthalpy and Kirchhoff transform method. *Transactions of the American Society of Agricultural Engineers*, **44**, 429–438.

Schwartzberg, H.G. (1976) Effective heat capacities for the freezing and thawing of food. *Journal of Food Science*, **41**, 152–156.

Schwartzberg, H.G., Singh, R.P., Sarkar, A. (2007) Freezing and thawing of foods – computation methods and thermal properties correlation. In Heat Transfer in Food Processing: Recent Developments in Food Processing, S. Yanniotis and & B. Sunden (Eds.), pp. 61–100. WIT Press, Boston.

Segerlind, L. (1984) *Applied Finite Element Analysis*, 2nd edn, John Wiley and Sons, New York.

Sekerka, R.F. (2004) Morphology: from sharp interface to phase field models. *Journal of Crystal Growth*, **264**, 530–540.

Sekerka, R.F., Wang, S-L (1999) Moving phase boundary problems, in *Lectures on the Theory of Phase Transformations*, 2nd Edition, H.I. Aaronson (Ed.), TMS, Warrendale, pp. 231–284.

Sethian, J.A. (2001) Evolution, implementation, and application of level set and fast marching methods for advancing fronts. *Journal of Computational Physics*, **169**, 503–555.

Sethian, J.A., Strain, J. (1992) Crystal growth and dendritic solidification. *Journal of Computational Physics*, **98**, 231–253.

Shi, X., Datta, A.K., Mukherjee, Y. (1998) Thermal stresses from large volumetric expansion during freezing of biomaterials. *Transactions of the ASME*, **120**, 720–726.

Shi, X., Datta, A.K., Mukherjee, Y. (1999) Thermal fracture in a biomaterial during rapid freezing. *Journal of Thermal Stresses*, **22**, 275–292.

Simatos, D., Champion, D., Lorient, D., Loupiac, C., Roudaut, G. (2009) Water in Dairy Products, in *Advanced Dairy Chemistry: Volume 3: Lactose, Water, Salts and Minor Constituents*, ed. P.F. Fox & P.L.H. McSweeney, pp. 457–526. Springer, New York.

Sinha, N.K. (2011) *Handbook of Vegetables and Vegetable Processing*. Blackwell, New York.

Smith, R.E., Nelson, G.L., Henrickson, R.L. (1968) Applications of gemoetry analysis of anomalous shapes to problems in transient heat transfer. *Transactions of the ASAE*, **14**, 44–47,51.

Sola-Morales, M.E., Valerio-Junco, L., Lopez-Malo, A., Garcia, H.S. (2010) Dielectric properties of foods: reported data in the 21st century and their potential applications. LWT Food Science and Technology, **43**, 1169–1179.

Sotani, T., Arabas, J., Kubota, H., Kijima, M. (2000) Volumetric behaviour of water under high pressure at subzero temperature. *High Temperatures – High Pressures*, **32**, 433–440.

Sussman, M., Smereka, P., Osher, S. (1994) A Level Set Approach for Computing Solutions to Incompressible Two-Phase Flow. *Journal of Computational Physics*, **114**, 146–159.

Tan, L. & Zabaras, N. (2006) A level set simulation of dendritic solidification with combined features of front-tracking and fixed-domain methods. *Journal of Computational Physics*, **211**, 36–63.

Tarnawski V.R., Cleland D.J., Corasaniti S., Gori F., Mascheroni R.H. (2005) Extension of soil thermal conductivity models to frozen meats with low and high fat content. International Journal of Refrigeration, **28**, 840–850.

Tchigeov, G. (1956). *Problems of the Theory of Food Freezing.* Food Industry, Moscow (in Russian).

Tchigeov G. (1979). *Thermophysical Processes in Food Refrigeration Technology.* Food Industry, Moscow (in Russian).

Thompson, J.F., Warsi, Z.U.A., Mastin, C.W. (1982) Boundary-fitted coordinate systems for numerical solution of partial differential equations—A review. *Journal of Computational Physics*, **47**, 1–108.

Tocci AM, Mascheroni RH (1994) Freezing times of meat balls in belt freezers: experimental determination and prediction by different methods. *International Journal of Refrigeration*, **17**, 445–452.

Toner, M., Cravalho, E.G., Karel, M. (1990) Thermodynamics and kinetics of intracellular ice formation during freezing of biological cells. *Journal of Applied Physics*, **67**, 1582–1593.

Toner, M. (1993) Nucleation of ice crystals inside biological cells, in *Advances in Low Temperature Biology* (ed. P. Steponkus), JAI Press, London, pp. 1–52.

Tremeac, B., Datta, A.K., Hayert, M., Le Bail, A. (2007) Thermal stresses during freezing of a two-layer food. *International Journal of Refrigeration*, **30**, 958–969.

Trujillo, F.J. (2004) A Computational Fluid Dynamic Model of Heat and Moisture Transfer during Beef Chilling, Ph.D. Thesis, School of Chemical Engineering and Industrial Chemistry, University of New South Wales, Sydney, Australia.

Udaykumar, H.S., Mittal, R., Shyy, W. (1999) Solid-liquid phase front computations in the sharp interface limit on fixed grids. *Journal of Computational Physics*, **153**, 535–574.

Udaykumar, H.S., Mao, L., Mittal, R. (2002) A finite-volume sharp interface scheme for dendritic growth simulations: comparison with microscopic solvability theory. *Numerical Heat Transfer, Part B*, **42**, 389–409.

van der Sluis, S.M. (1993) Cooling and freezing simulation of bakery products, *Proceedings of International Institute of Refrigeration Meeting Commissions B1, B2, D1, D2/3*, Palmerston North, New Zealand, 1993.

van der Sman, RGM (2008) Prediction of enthalpy and thermal conductivity of frozen meat and fish products from composition data. *Journal of Food Engineering*, **84**, 400–412.

van der Sman, RGM, Boer, E (2005) Predicting initial freezing point and water activity of meat products from composition data. *Journal of Food Engineering*, **66**, 469–475.

Voller, V.R. (1990) A fast implicit finite difference method for the analysis of phase change problems. *Numerical Heat Transfer B*, **17**, 155–169.

Voller, V.R. (1996) An overview of numerical methods for solving phase change problems, in *Advances in Numerical Heat Transfer* (eds W.J. Minkowycz and E.M. Sparrow), Taylor & Francis, London, Vol. 1, pp. 341–375.

Voller, V.R. (2008) An enthalpy method for modeling dendritic growth in a binary alloy. *International Journal of Heat and Mass Transfer*, **51**, 823–834.

Voller, V.R., Cross, M. (1985) Accurate solutions of moving boundary problems using the enthalpy method. *International Journal of Heat and Mass Transfer*, **24**, 545–556.

Voller, V.R., Prakash, C. (1987) A Fixed-Grid Numerical Modeling Methodology for Convection-Diffusion Mushy Region Phase-Change Problems. *International Journal of Heat and Mass Transfer*, **30**, 1709–1720.

Voller, V.R., Swaminathan, C. R. (1991) Generalized Source-Based Method for Solidification Phase Change. *Numerical Heat Transfer Part B*, **19**, 175–189.

Voller, V.R., Swaminathan, C.R., Thomas, B.G. (1990) Fixed grid techniques for phase change problems: a review. *International Journal of Numerical Methods in Engineering*, **30**, 875–898, 1990.

Wagner, W., Pruss, A. (2002) The IAPWS formulation 1995 for the thermodynamic properties of ordinary water substance for general and scientific use. *Journal of Physical and Chemical Reference Data*, **31**, 387–535.

Wagner, W., Riethmann, T., Feistel, R., Harvey, A.H. (2011) New equations for the sublimation pressure and melting pressure of H2O ice Ih. *Journal of Physical and Chemical Reference Data*, **40**, 043103.

Wang J.F., Carson J.K., North M.F., Cleland D.J. (2006) A new approach to modelling the effective thermal conductivity of heterogeneous materials. *International Journal of Heat and Mass Transfer*, **49**, 3075–3083.

Wang J.F., Carson J.K., North M.F., Cleland D.J. (2008) A new structural model of effective thermal conductivity for heterogeneous materials with co-continuous phases. *International Journal of Heat and Mass Transfer*, **51**, 2389–2397.

Wang J.F., Carson J.K., Willix J., North M.F., Cleland D.J. (2010) Prediction of thermal conductivity for frozen foods with air voids. *1st IIR International Cold Chain Conference*, Cambridge UK, 2010.

Wheeler, A.A., Boettinger, W.S., McFadden, G.B. (1992) Phase-field model for isothermal phase transitions in binary alloys. *Physical Review A*, **45**, 7424–7440.

Woinet, B., Andrieu, J., Laurent, M. (1998a) Experimental and theoretical study of model food freezing, Part 1: Heat transfer modelling. *Journal of Food Engineering*, **35**, 381–93.

Woinet, B., Andrieu, J., Laurent, M., Min, S.G. (1998b) Experimental and theoretical study of model food freezing, Part 2: Characterization and modelling of ice crystal size. *Journal of Food Engineering*, **35**, 394–407.

Ye, T., Mittal, R., Udaykumar, H.S., Shyy, W. (1999) An accurate cartesian grid method for viscous incompressible flows with complex immersed boundaries. *Journal of Computational Physics*, **156**, 209–240.

Zhang, H., Datta, A.K. (2000) Coupled electromagnetic and thermal modeling of microwave oven heating of foods. *Journal of Microwave Power & Electromagnetic Energy*, **35**, 71–85.

Zienkiewicz, O.C., Taylor, R.L. (1991) *The Finite Element Method*, McGraw-Hill, London.

Zorrilla, S.E., Rubiolo, A.C. (2005) Mathematical modeling for immersion chilling and freezing of foods. Part I: Model development. *Journal of Food Engineering*, **66**, 329–338.

Index

A
Alternating direction algorithm, 76
Analytical methods, 4
Apparent specific heat, 8, 16, 17, 65, 83, 84, 86, 89
Apparent specific heat method, 17, 83, 89
Approximate methods, 4, 15, 37, 38
Arroyo, J.G., 54

B
Banaszek, J., 75
Bart, G.C.J., 62
Bathe, K.J., 83
Becker, B.R., 6, 21, 44, 45
Biot number, 6, 29, 30, 41, 43, 46, 48, 52, 61, 133
Bird, R.B., 70
Boer, E., 10
Bonacina, C., 22
Botte, G.G., 69, 71
Boundary condition, 26, 65, 67–70, 73, 92
Bound water, 9, 11
Budhia, H., 32
Bulk modulus, 18

C
Cahn, J.W., 35
Califano, A.N., 76
Calvelo, A., 38, 40, 44
Capacitance matrix, 69, 79
Carslaw, H.S., 25, 32, 38, 61
Castaigne, F., 39, 44
Cellular automata, 134
Choi, Y., 7, 8, 14, 15, 18, 19, 24
Cleland, A.C., 19, 22, 26, 37, 42, 44–46, 48, 59, 61, 81, 84
Cleland, D.J., 48, 49, 51, 55–57
Cleland & Earle's method, 45, 46

Comini

Comini, G., 22, 84, 86, 88, 89
Computational fluid dynamics (CFD), 6, 134
Conductance matrix, 69
Control volume, 69, 70, 81, 83, 90
Convergence of numerical methods, 76, 91, 92
Crank, J., 77–79, 85
Crank-Nicolson method, 77, 79
Crystal growth, 2, 134

D
Datta, A.K., 5
Davey, L.M., 6
De Michelis, A., 44
Dendritic growth, 90
Density, 5–7, 17–19, 26, 27, 90, 91
Dirichlet boundary condition, 30, 32, 66
Discretization of space, 4
Domoto, A., 35
Douglas, J. Jr., 80
Duckworth, R.B., 10
Dynamic heat load, 58, 59, 61–63

E
Earle, R.L., 22, 48, 61, 81
Effective medium theory (EMT) model, 19–21
Embedded mesh, 90
Empirical methods, 4, 37, 44, 45
Enthalpy, 5, 11–16, 18, 22, 23, 38, 40, 46, 56, 58, 61, 79, 82–84, 86, 88, 89, 92, 133
Enthalpy method, 84–88, 133
Equivalent heat transfer dimensionality (EHTD), 48, 50–52
Equivalent shape, 48
Eulerian mesh, 90
Euler method, 77, 79
Evaporative cooling, 6
Eyres, N.R., 85

Q. T. Pham, *Food Freezing and Thawing Calculations,*
SpringerBriefs in Food, Health, and Nutrition, DOI 10.1007/978-1-4939-0557-7,
© The Author 2014